Laboratory Manual
to accompany

CONTEMPORARY ELECTRIC CIRCUITS
Insights and Analysis

Second Edition

Richard J. Lokken
Milwaukee Area Technical College

Robert A. Strangeway
Milwaukee School of Engineering

John D. Gassert
Milwaukee School of Engineering

Owe G. Petersen
Milwaukee School of Engineering

PEARSON

Prentice
Hall

Upper Saddle River, New Jersey
Columbus, Ohio

Acquisitions Editor: Kate Linsner
Production Editor: Rex Davidson
Design Coordinator: Diane Ernsberger
Editorial Assistant: Lara Dimmick
Cover Designer: Thomas Mack
Cover art: Getty One
Production Manager: Matt Ottenweller
Marketing Manager: Ben Leonard

This book was set by Richard J. Lokken and was printed and bound by Banta Book Group. The cover was printed by Coral Graphic Services, Inc.

Pearson Education Ltd.
Pearson Education Singapore Pte. Ltd.
Pearson Education Canada, Ltd.
Pearson Education—Japan

Pearson Education Australia Pty. Limited
Pearson Education North Asia Ltd.
Pearson Educación de Mexico, S.A. de C.V.
Pearson Education Malaysia Pte. Ltd.

10 9 8 7 6 5 4 3 2 1

ISBN: 0-13-111560-X

Preface

Students, Please Read!

This laboratory manual can be used in whole or in part to supplement the textbook *Contemporary Electric Circuits: Insights and Analysis,* Second Edition for technician, engineering technician, or engineering technologist programs.

The format for each experiment is subdivided as follows. Learning objectives are provided. The background section provides abbreviated theory needed for the laboratory experiment. The detailed theory is provided in the textbook *Contemporary Electric Circuits: Insights and Analysis,* Second Edition. A parts list and a procedure are provided for each experiment. Some numbering of steps has been incorporated for reference purposes, but each individual step is not numbered. The questions at the end of the laboratory experiment serve to guide you in writing the Analysis of Results and the Conclusion sections of your laboratory reports.

The experimental documentation starts as a "fill-in-the-blank" form. After Experiment 5, the documentation format utilizes the industry standard engineering notebook (typical grid pages are included in this manual). You will write the purpose in your own words in the Introduction section in your engineering notebook prior to the laboratory session. The learning objectives are *not* the purpose that you should write in the introduction. You will also perform a theoretical analysis on each circuit prior to the beginning of the laboratory session. You will record the calculations for this analysis in the Theoretical Solution section in your engineering notebook. The calculated results from this analysis will be recorded in a results table in the Results section in your engineering notebook. The measured data will be recorded and compared to the theoretical data. After completing the laboratory session, you will write an Analysis of Results section and a Conclusion section, both in your engineering notebook.

The basic test equipment required for most sessions are a three-output DC power supply, DMM, VOM, oscilloscope, and a signal generator.

The PSpice® simulation outputs seen in Appendix B are used with permission of Cadence Design Systems, Inc. Cadence, the Cadence logo, OrCAD, OrCAD Capture, and PSpice are registered trademarks of Cadence Design Systems, Inc.

The authors wish to thank the students of Milwaukee Area Technical College and Milwaukee School of Engineering for their helpful suggestions in writing this lab manual. The authors also wish to thank our colleagues and administration for the support and guidance we have received in this project.

Table of Contents

Experiment 1: INTRODUCTION TO THE ELECTRICAL LABORATORY

Learning Objectives

After completing this experiment, you should be able to:

- Identify electrical laboratory safety rules and situations to which they apply.
- Explain why engineering notebooks are important to engineers, engineering technologists, engineering technicians, and technicians.
- Identify important engineering notebook practices.
- Identify experiment write-up sections for use in an engineering notebook and/or a laboratory report and explain the organization of those sections.

Background: Electrical Laboratory Safety Rules[1]

1. Do *not* rely on fuses, relays, circuit breakers, or interlock devices for your safety.

2. When in doubt, do *not* turn the power ON until you have checked the circuit with your instructor.

3. Do *not* work on wet floors.

4. Do *not* work alone in laboratories (minimum of two students).

5. Turn OFF power to all equipment when finished.

6. Make all circuit connections before applying power, that is, make the connection to the power source (turned OFF) your last connection to complete the circuit. Then turn ON the source. Tear down the circuit in the reverse order by first turning OFF the source and then disconnecting it.

7. Do *not* change circuit components with the power ON. Turn the power OFF. Besides the danger to yourself, you could damage the components if you change them with the power ON.

8. *CAUTION*: Many electrical components are *very hot* when operating.

9. Pay particular attention to the *polarity* of polarized capacitors.

---

[1] The rules stated here do *not* form a complete list of safety rules. They are simply general guidelines to electrical safety. Your instructor should provide you with specific safety rules appropriate to your laboratory environment.

10. Know where the main power and/or emergency power down switches are located in the laboratory and how to use them.

11. Do *not* use any defective equipment or components. Report such occurrences to the instructor *immediately* upon discovery.

12. Report *any* injuries to the instructor *immediately*.

13. Do *not* bypass the safety feature of a three-pronged plug. If the socket can accept two-pronged plugs only, use an adapter and be sure to connect the ground lead to a good ground (*not* the center screw that holds the cover on) *before* you insert the plug into the socket. Always check for proper grounding beforehand!

14. In the event of an electrical fire:
    If possible without danger to yourself,
    a. De-energize the circuit (turn OFF the power)
    b. Notify the instructor as quickly as possible
    c. Instructor will use an extinguisher that can put out an electrical fire
    d. Remove all other possible fuels, such as paper
    e. Administer first aid to anyone who needs it
    f. Call the appropriate safety department, if necessary (Pull fire alarm, evacuate the building, and call from someplace outside the building, if appropriate.)

15. If a component burns, *avoid* breathing the fumes. The fumes could be toxic and poisonous.

16. Do *not* wear loose or flappy clothes around moving machinery (such as a motor). Also, be sure to wear goggles.

17. Use only one hand at a time in any high voltage electrical circuit to prevent putting yourself in the circuit.

18. Do *not* attempt to directly pull someone from a "live" circuit. Turn the power OFF if possible. Otherwise, remove the person with a non-conducting object.

Background: Engineering Notebook Practices

An engineering notebook is a legal document used by engineers, technologists, and technicians to record all work performed. Patents have been awarded based on the information in an engineering notebook. The following engineering notebook practices are essential:

1.	The engineering notebook must be bound with numbered pages. *Never* remove a page from an engineering notebook.

2.	Your name, company, and (if appropriate) title of the work or project *must* be on the front cover.

3.	All entries should be made in *black* ink (for copying purposes).

4.	Entries are to be detailed descriptions of the work performed, the equipment and materials used, the theoretical and measured results of the experiment, and so on. Enter all work as it is performed.

5.	All entries are to be dated and signed by the person performing the work. All pages *must* be dated and signed at the bottom.

6.	An associate who witnessed the work should attest to the entry by co-signing the engineering notebook.

7.	*Never* "BLOT OUT" or erase an error. Cross it out with a single line then write your initials and the date above the crossed out text. If a whole section is in error, cross it out and give the reason it is in error in the margin.

8.	Copies and other information that are to be included in the engineering notebook must be permanently attached to notebook pages.

9.	The logbook is used as a reference when preparing laboratory reports.

A typical organization format follows next. Appendix A contains a sample lab report. Appendix B contains a sample engineering notebook entry.

Background: Engineering Notebook and Laboratory Report Organization

There are various methods of organization that are used in engineering notebooks and laboratory reports. The following format lists the sections that are commonly used (perhaps with different titles). Learn one format well; that is, learn and understand the logic behind the organization of report writing. You will then be well equipped to adapt the organization to the needs of future courses and eventual company requirements. Be sure to write in *third person*.

1.	Title of Experiment*
	a.	entry at the top of a new page in the engineering notebook
	b.	title page and contents page in a formal laboratory report

2.	Introduction*
	a.	statement of the problem to be investigated

b. any significant assumptions to be made
c. concise statement that establishes the *purpose* of this experiment

3. Theoretical Solution*
   a. circuit diagram(s)
   b. *theoretical* calculations
   c. theoretical simulation solutions

4. Results
   a. tables and spreadsheet tables*
      (usually show theoretical results and measured results side-by-side)
   b. measurement procedure (usually a short summary in paragraph form; enter any special measurement diagrams)
   c. simulation solutions using *actual* component values
   d. any other results, such as plots generated from a measurement instrument
   e. samples of calculations based on *measured* results

5. Analysis of Results
   a. other displays of results, such as graphs generated from the measured data
   b. discussion (analysis) of results
   c. error analysis
   d. any other appropriate analysis

6. Conclusion
   a. concise statements of what was proven, shown, validated, verified, discovered, and so on during the experiment
   b. any critical results along with their significance
   c. explanation and evaluation of meeting the purpose of the experiment as stated in the Introduction section

* Students *must* complete these items and enter them into their engineering notebooks *before* the scheduled laboratory session. See Appendix A for a sample lab report and Appendix B for a sample engineering notebook.

Experiment 2: THE NATURE OF ELECTRICITY

Learning Objectives

After completing this laboratory experiment, you should be able to:

- Describe what electricity is and how electricity is used.
- Describe the functions of sources, conductors, and loads.
- Describe what is needed to form a complete circuit.
- Describe how energy is converted at both the source and the load.

Background: What Is Electricity?

Electricity is a form of energy. Energy is the ability to do work, such as heat liquid or move objects. A key physical law is the conservation of energy. This states that energy can be neither created nor destroyed but may be transferred or converted from one form into another. Therefore, we will examine how energy is converted.

A battery is a typical source of electrical energy in today's mobile world. But a battery does *not* provide electrical energy directly. The source of energy in a battery is chemical. The form of the output energy is electrical. We will also use a solar cell, which converts light energy into electrical energy, and a generator, which converts mechanical energy into electrical energy.

To utilize energy, the energy must transfer from the source to the load. Loads can be light bulbs, motors, heaters, fans, and so on. The load is what we are trying to operate. The transfer of energy takes place using conductors or wires.

Prelaboratory Preparation

For each circuit in this experiment, predict the input form of energy and the output form of energy in both the source and the load.

Table 2-1 Energy conversion results

Figure	Component	Predicted Energy Conversion	
		From	To
2-2	3 V battery	Chemical	Electrical
2-2	0.75 W bulb	Electrical	Light
2-3	3 V battery		
2-3	3 V motor		
2-4	3 V battery		
2-4	39 Ω resistor		
2-5	Solar cell		
2-5	LED		
2-6	Generator		
2-6	0.75 W bulb		

"Can energy be transferred through the air?" Think carefully about this and give examples to support your answer.

_____

_____

_____

_____

Parts List

- Battery Holder
- 3 V DC motor
- 3 V 0.75 W light bulb
- (Qty. 2) 1.5 V batteries
- 39 Ω, 5 %,1/4 watt resistor
- Solar cell capable of 3 VDC output
- Small fan blade
- 3-foot wooden dowel ½ inch diameter
- Handle to operate the DC motor as a generator
- Proto board
- Red LED

NOTE: The DC motor, light bulb, LED, and solar cell can be obtained from Electronix Express (www.elexp.com.)

Procedure

A proto board, also called a breadboard, is used in most experiments to construct the circuit. A typical proto board is shown in Figure 2-1.

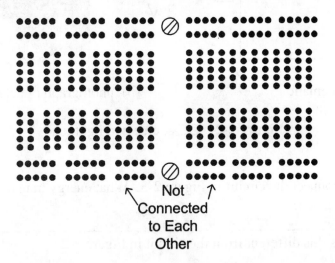

Figure 2-1 Typical proto board

1. The connections are made in columns by groups of five for this proto board. To check the board, an instrument called an ohmmeter can be used to determine which groups are connected. Your instructor will direct you in the use of an ohmmeter.

2. The circuit in Figure 2-2 shows a DC battery, a switch, and a load. Connect this circuit using a proto board and comment on what is taking place when the switch is closed. What energy conversions are taking place?

_____

_____

What happens when the switch is opened?

_____

_____

_____

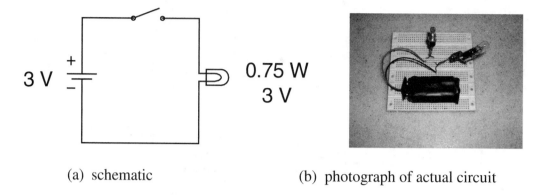

(a) schematic   (b) photograph of actual circuit

Figure 2-2   DC source and light load

3. Now connect the circuit in Figure 2-3. What energy transformations are taking place?

_____

How is this different from the circuit in Figure 2-2?

_____

_____

.

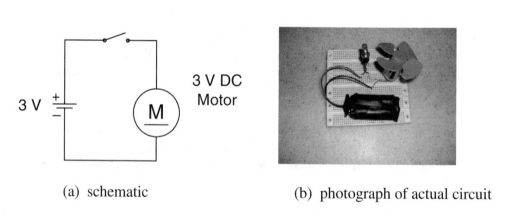

(a) schematic   (b) photograph of actual circuit

Figure 2-3   DC source and DC motor load with a fan

4. Disconnect the fan from the motor. Reconnect the fan with the addition of a 3-foot wooden dowel shaft between the motor and the fan. What happens to the speed of the fan?

_____

_____

_____

5. Remove the wooden dowel shaft and reconnect the fan directly to the motor. Add 3 feet of wire between the source and the motor. What happened to the fan speed relative to the setup with the 3-foot dowel shaft?

_____

_____

6. The next circuit uses a resistor for the load. Resistors will be discussed in greater detail in later experiments. One thing a resistor does is dissipate energy. It does this by converting electrical energy into heat. This resistor is sized such that the power dissipated is below the power rating of the resistor, but the resistor will get warm. Build the circuit in Figure 2-4. Close the switch and touch the resistor. Be careful, the resistor may be warm to the touch. It may even be hot. Open the switch once you feel the resistor heat up. This heating is due to the energy conversion taking place. What are some possible uses for this type of circuit?

_____

_____

_____

_____

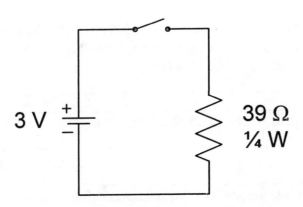

Figure 2-4 DC source and a resistor load

7. The first three circuits used a battery as the source. There are other ways to obtain or generate electrical energy. One such device is a solar cell. Now connect the solar cell to an LED, as shown in Figure 2-5. Ensure that light is shining on the cell and close the switch. What happens?

_____

_____

Now cover the solar cell. What happens?

_____

_____

Why?

_____

What energy conversions are taking place?

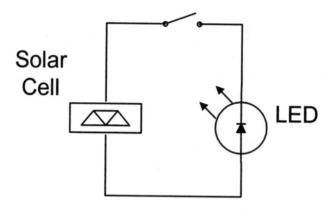

Figure 2-5 Solar cell and LED circuit

8. Connect the circuit shown in Figure 2-6. Attach a hand crank to the shaft
   of the motor. Turn the crank and observe what happens. What happens
   when the crank is turned slowly?

What happens when the crank is turned faster?

How does the operation of the generator differ from the operation of the
motor?

What energy conversions are taking place?

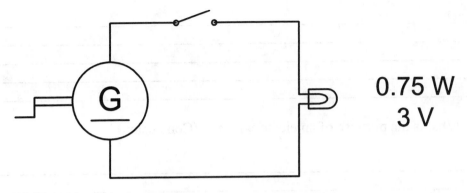

0.75 W
3 V

Figure 2-6 Generator circuit

Questions

Note to the student: These questions serve to guide you in the writing of your
Analysis of Results and Conclusion sections in your
engineering notebook. Usually, you must decide which
section your answer belongs in. For the questions in this
experiment, the sections to which your responses belong
are indicated to illustrate this practice.

1. In one table, list the energy sources used in each circuit of this experiment and
    what the energy conversion was in each source. Compare to your predictions.
    What is common to all sources? (Analysis)

_____

_____

_____

_____

_____

2. What is the purpose of an electrical source? (Conclusion)

_____

_____

_____

_____

_____

3. In another table, list the loads used in each circuit of this experiment and what
    energy conversion took place in each load. Compare to your predictions.
    What is common to all loads? (Analysis)

11

_____
_____
_____
_____
_____

4. What is the purpose of an electrical load? (Conclusion)

_____
_____
_____
_____
_____

5. How did the energy transfer from the source to the load in all circuits? (Conclusion)

_____
_____
_____
_____
_____

6. What was the purpose of the switch? (Conclusion)

_____
_____
_____
_____
_____

7. Which form of energy is transported over the 3 feet in the setup that uses the battery, electric motor, 3-foot dowel, and fan? (Analysis)

_____
_____
_____
_____
_____

8. What happened when the motor was "loaded" with the wooden dowel relative to the setup without the dowel? (Analysis)

_____
_____
_____

_____
_____

9. Which form of energy is transported over the 3 feet in the setup that uses the battery, 3 feet of wire, electric motor, and fan? (Analysis)

_____
_____
_____
_____
_____

10. Which form of energy is more efficient to transport over distance? Why? (Conclusion)

_____
_____
_____
_____
_____

11. What energy conversion took place when the motor was used as a generator in the circuit shown in Figure 2-6? (Analysis)

_____
_____
_____
_____
_____

12. Can a motor be used as both a source and a load? In general, can every component be used as both a source and a load? Explain. (Conclusion)

_____
_____
_____
_____
_____

Experiment 3: Electrical Measurements

Learning Objectives

After completing this laboratory experiment, you should be able to:

- Perform voltage and current measurements.
- Calculate power from voltage and current measurements.
- Calculate resistance from voltage and current measurements.

Background: Making Electrical Measurements

There are two general-purpose types of meters used to measure voltage, current, and resistance. The digital multimeter (DMM) is the most common general-purpose meter. The DMM has a convenient digital readout. This meter auto scales (adjusts itself to the proper scale) based on the measurement selected. However, if the quantity being measured is changing, it is hard to see a trend (flashing numbers are hard to interpret), whereas it is easier to see the trend in a moving needle on an analog meter.

The volt-ohm-meter (VOM) is an analog meter that has a needle that rotates on a curved scale. Although not as commonly used as the DMM, it is still utilized in many applications. The discussion that follows is generic to both meter types, but additional scale-related notes are included for the VOM because the VOM does *not* auto-scale.

Review the operator manuals for the meters in your laboratory. Ask your instructor to obtain the manuals, if necessary.

The Voltmeter

Use the voltage scale to measure voltages. The meter is now being used as a voltmeter. Be sure to check what kind of voltage you are measuring (AC or DC).

VOM note: Set the voltmeter on a range higher than the voltage that you are measuring (e.g., set the voltmeter on the 250 V range to measure 130 V). Use the highest range on the voltmeter if you do not know how large the voltage value may be.

Always connect the voltmeter *across* the device whose voltage you want to measure, as shown in Figure 3-1.

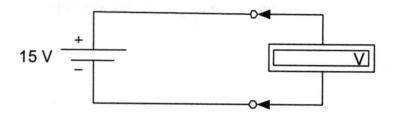

Figure 3-1 Voltage measurement with a voltmeter
(Arrows indicate meter lead connections)

As another example, let's say that there is resistor in a circuit and you want to know the voltage across it. See the diagram in Figure 3-2. Note that the voltmeter is connected *across* the resistor. Again, always connect the voltmeter *across* the device whose voltage you want to measure.

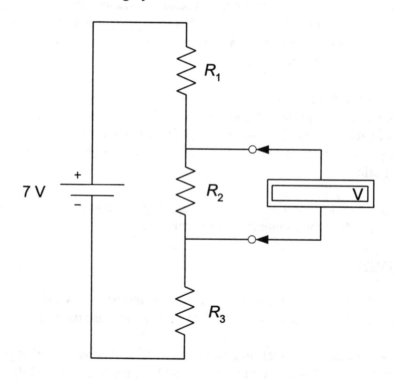

Figure 3-2 Voltage measurement across a resistor

The Ammeter

Use the current scale to measure DC currents. The meter is now being used as an ammeter. Be sure to check what kind of current you are measuring (AC or DC).

VOM note: Set the ammeter on a range higher than the current that you are measuring (e.g., set the ammeter on the 100 mA range to measure 90 mA). Use the highest range on the ammeter if you do not know what the current may be.

Always connect the ammeter in series with (in the path of) the circuit components. For example, to measure the current in the circuit shown in Figure 3-3, you want to find the current in the path of the electricity; therefore, you put the ammeter in series.

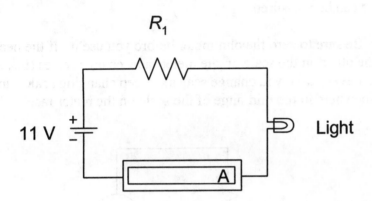

Figure 3-3 Current measurement with an ammeter

NEVER put an ammeter in parallel (across) any device, such as shown in Figure 3-4. You could damage the ammeter.

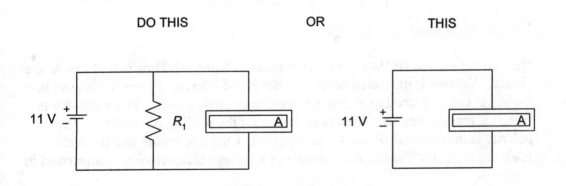

and the result: POOF!

Figure 3-4 Incorrect and dangerous ammeter connection

17

The Ohmmeter

Use the resistance scale to measure resistances. The meter is now being used as an ohmmeter. **Never put an ohmmeter into any circuit that has the power turned on.** You could damage the ohmmeter.

Put the ohmmeter leads across the device whose resistance you want to measure, as shown in Figure 3-5. The device needs to be removed from the circuit before the resistance can be measured.

VOM notes: Be sure to zero the ohmmeter before you use it. If the needle is at one end or the other on the scale on the meter face, change scales (be sure to re-zero the meter every time you change scales). Keep changing scales until the needle is somewhere in the midrange of the scale on the meter face.

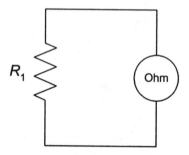

Figure 3-5 Resistance measurement with an ohmmeter

Making measurements in circuits

How do we use the DMM (or VOM) to measure voltage? How is it placed in the circuit? Voltage is measured across the device of interest. Figure 3-6 shows how the meter can be placed to measure the voltage across a lamp. The voltmeter is placed across the device. What does it mean if the reading is negative? The polarity is the opposite of what was expected. Check to ensure that the meter leads are correctly placed. An example of a voltage measurement is illustrated in Figure 3-7.

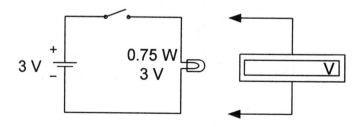

Figure 3-6 Lamp test circuit

18

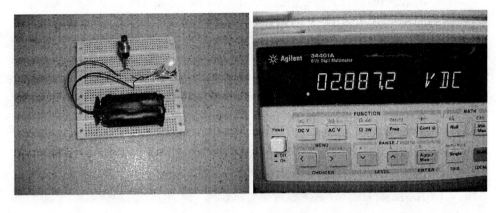

(a) lamp on   (b) voltage meter reading

Figure 3-7   The voltage reading on the DMM with the lamp on
(Reprinted with permission of Agilent)

If current is the desired measurement then a different connection method is needed. The current meter needs to be placed *in* the circuit so that the current flows *through* the ammeter. Break the current path and place the ammeter where the circuit was broken. See Figures 3-8 and 3-9.

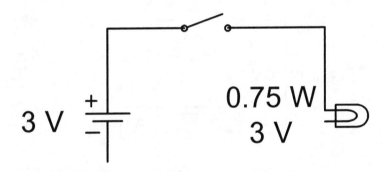

Figure 3-8   The current path has been broken

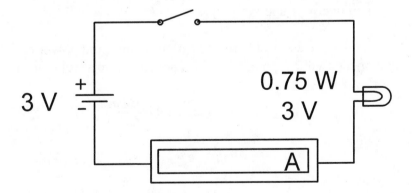

Figure 3-9   The ammeter is placed in series in the circuit

Now the current can be measured. An example reading is shown in Figure 3-10.

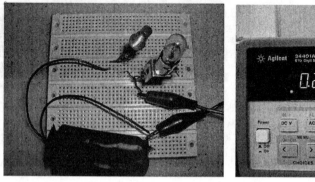

(a) meter connection   (b) meter reading

Figure 3-10   Current reading setup and result
(Reprinted with permission of Agilent)

What if both current and voltage were wanted at the same time? Two meters
would be needed. Place the meters as shown in Figure 3-11.

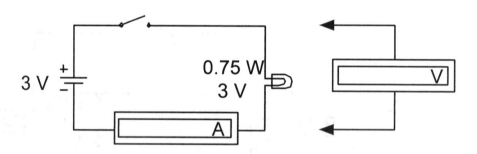

Figure 3-11   Both current and voltage measurements

Calculating power and resistance

Once voltage and current are known, resistance and power can be calculated. The
fundamental voltage, current, resistance, and power relationships are

$$Resistance = R = \frac{Voltage}{Current} = \frac{V}{I}$$

$$Power = P = (Voltage)(Current) = VI$$

For the circuit in Figure 3-11, if the voltage across the bulb is 3 volts and the
current through the bulb is 250 mA, then the resistance and power of the bulb
are:

$$R = I = \frac{3 \text{ V}}{250 \text{ mA}} = 12 \ \Omega$$

$$P = VI = (3 \text{ V})(250 \text{ mA}) = 750 \text{ mW}$$

Prelaboratory Preparation

Create a results table in your engineering notebook to record measured voltage and current for each test circuit. Columns for experimental resistance and power are also desired. An example table format is shown in Table 3-1.

Table 3-1 Sample results table

Test Circuit Number	Measured Voltage	Measured Current	Experimental Resistance	Experimental Power
Test Circuit 1				
Test Circuit 2				
Test Circuit 3				
Test Circuit 4				
Test Circuit 5				

Parts List

- Battery Holder
- 3 V DC motor
- 3 V light bulb
- (Qty. 2) 1.5 V batteries
- 33 Ω, 5%, 1/4 watt resistor
- Solar cell capable of 3 VDC output
- Small fan blade
- 3 foot wooden dowel ½ inch diameter
- Handle to operate the DC motor as a generator
- Proto board
- Red LED

Procedure

Voltage and current are measured for each circuit presented in Experiment 2 (see Figures 3-12 through 3-16). The resistance and the power used can be calculated from the voltage and current measurements. Record all the values in the table you entered previously in the Results section of your engineering notebook.

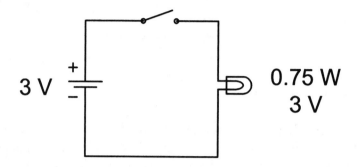

Figure 3-12   Test circuit 1

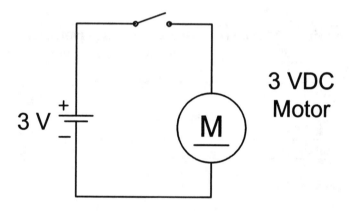

Figure 3-13   Test circuit 2

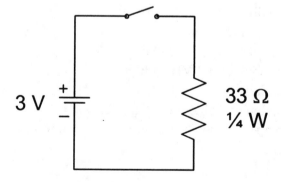

Figure 3-14   Test circuit 3

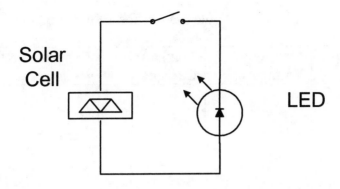

Solar
Cell

LED

Figure 3-15 Test circuit 4

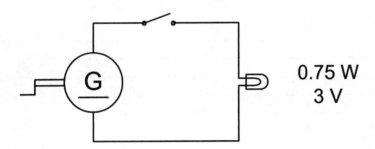

G

0.75 W
3 V

Figure 3-16 Test circuit 5

What is the maximum voltage that can be generated in test circuit 5? Record your observation in the Results section.

Table 3-2 Results Table One

Test Circuit Number	Measured Voltage	Measured Current	Experimental Resistance	Experimental Power
Test Circuit 1				
Test Circuit 2				
Test Circuit 3				
Test Circuit 4				
Test Circuit 5				

Table 3-3 Results Table Two

Test Circuit	Maximum Voltage	Maximum Current	Experimental Resistance	Maximum Power
Test Circuit 5				

Questions

1. Compare the voltage measurements for the test circuits. (Analysis)

2. How did the voltage measurements compare to the supply voltage? Explain. (Analysis)

3. What was the experimental power in the test circuit in Figure 3-12? How did that compare to the rated power of the lamp? (Analysis)

4. Was the generator able to provide enough power to light the lamp? Explain. (Conclusion)

5. Would a solar cell be an efficient energy generator in a classroom environment? Explain. (Conclusion)

_____

_____

_____

_____

_____

6. Calculate the percentage error for resistance in the circuit shown in Figure 3-14. (Analysis)

$$\% \ error = \frac{\text{measured } R - \text{nominal } R}{\text{nominal } R} \times 100\%$$

Experiment 4: Resistance of Resistors and Wires

Learning Objectives

After completing this laboratory experiment, you should be able to:

- Determine the nominal resistance value, tolerance, and range of a resistor using the resistor color code.
- Measure the resistance of various resistors and potentiometers using an ohmmeter.
- Calculate the percentage error between the expected value and the measured value of resistance.
- Determine the resistivity of a conductor material from wire measurements.

Background

The resistor is just one component that has resistance. The loads used in the first two experiments have resistance. All components have some resistance.

Resistors limit the current in a circuit. A resistor is a special type of component that converts electrical energy to heat. The resistance value of resistors can be determined by the "color code" of the resistor. This color code is made up of color bands as shown in Figure 4-1. The various color bands on a resistor indicate its nominal resistance value and tolerance.

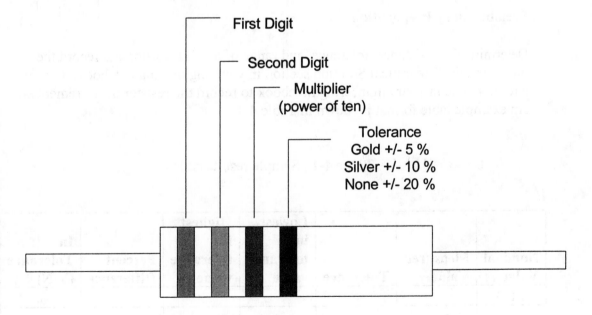

Figure 4-1   Resistor color code markings

The colors and corresponding values of the first three bands are:

Black	0		Green	5
Brown	1		Blue	6
Red	2		Violet	7
Orange	3		Gray	8
Yellow	4		White	9

The tolerance colors in the fourth band are:

Gold   5%   Silver 10%   None 20%

For example, brown—red—orange—gold = 12000 $\Omega \pm 5\%$, or 12 k$\Omega \pm 5\%$. Resistors are also characterized by the maximum power they can absorb, before the excessive heat destroys them. Common values used in the laboratory experiments are ¼ watt. Physically larger resistors can dissipate ½ watt, 1 watt, and so on. **The common 5% resistors** come only in certain values. The first two digits are: **10, 12, 15, 18, 22, 27, 33, 39, 47, 56, 68, and 82.**

There are several types of variable resistors. In this experiment you examine the operation of the potentiometer. The potentiometer can be used to allow the operator to change resistance levels. For example, a volume control on a radio could be a potentiometer and a potentiometer is used to vary the speed of an electric train.

Prelaboratory Preparation

Determine the resistance, tolerance, and range of a 1 kΩ resistor and record the process in the Theoretical Solution section in your engineering notebook. Create a results table in your engineering notebook to record the resistor measurements. An example table format is shown in Table 4-1.

Table 4-1   Sample results table

Nominal Value	Measured Value	Tolerance	Lowest in-tolerance value	Highest in-tolerance value	Percent Difference	In Tolerance (Y/N)
1 kΩ						

Parts List

- (Qty. 3) resistors
- 1 kΩ resistor
- potentiometer

Procedure

1. Obtain four resistors of different values from the instructor (one of which should be 1 kΩ). Read the color code to determine the resistance value including the tolerance. Measure all four resistors and record the results. Compare the measured results to the nominal results. Determine the percentage difference and determine if the resistors are in tolerance.

Nominal Value	Measured Value	Tolerance	Lowest in-tolerance value	Highest in-tolerance value	Percent Difference	In Tolerance (Y/N)
1 kΩ						

2. Now connect the resistors end-to-end (in series) as shown in Figure 4-2. Measure the total resistance of this combination.

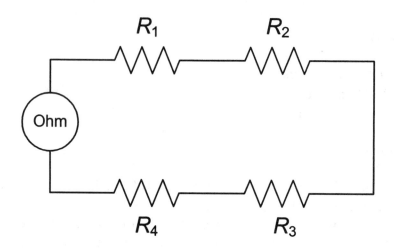

Figure 4-2   Resistor circuit

Sum of Nominal Value	Measured Value	Tolerance	Lowest in-tolerance value	Highest in-tolerance value	Percent Difference	In Tolerance (Y/N)

3. Obtain a potentiometer from the instructor. Measure the resistance across the two outside terminals. Then measure the resistance from the center terminal to one outside terminal and then to the other outside terminal. How do these two readings relate to the first reading? Next rotate the shaft of the potentiometer and repeat all three measurements. What changed? What stayed the same?

Shaft	Nominal Value	Outside Terminals Resistance	Center to Left Side Terminal	Center to Right Side Terminal
Fully CW				
Fully CCW				
Middle Position				

4. Obtain a wire sample from your instructor. Measure the diameter of the wire (conductor, not insulation) and its length. Measure the resistance of the wire with an ohmmeter that has sufficient sensitivity.

Wire Diameter	Wire Length	Wire Conductivity	Calculated Wire Resistance	Measured Wire Resistance

Questions

1. Were all the resistors in tolerance? Explain. (Analysis)

_____

_____

_____

_____

_____

2. Determine how the individual resistors would mathematically combine to obtain the measured resistance of the four resistors connected end-to-end. (Analysis)

_____

_____

_____

_____

_____

3. How does the potentiometer change resistance? Refer to step 3 of the procedure. (Conclusion)

_____

_____

_____

_____

_____

4. Name some common uses for a potentiometer. (Conclusion)

_____

_____

_____
_____
_____

5. Why would a rise in temperature damage a resistor? (Analysis)

_____
_____
_____
_____
_____

6. Calculate the resistivity of the wire sample that you measured. Determine the most probable conductor material from the experimental resistivity value. (Analysis)

_____
_____
_____
_____
_____

7. Discuss one method of determining the conductor material in an unknown wire. (Conclusion)

_____
_____
_____
_____
_____

Experiment 5: Series Circuits and Ohm's Law

Learning Objective

After completing this laboratory experiment, you should be able to:

- Demonstrate the validity of Ohm's law for a series circuit.

Background

Georg Simon Ohm identified the cause-and-effect relationships between voltage, current, and resistance. These relationships are stated in Ohm's law:

$$I = \frac{V}{R}$$

This equation shows that if the resistance is fixed, the current is directly proportional to the applied voltage. Also, if the voltage is fixed, the current is inversely proportional to the resistance. One can predict the behavior of the circuit based on this relationship.

Prelaboratory Preparation

Calculate the total resistance for the circuit shown in Figure 5-1. Then use Ohm's law to predict the total current. Use this information to predict the voltage across each resistor and the power dissipated by each resistor. Record all theoretical calculations in the Theoretical Solution section of your engineering notebook. Record your predictions in a table similar to Table 5-1 in the Results section of your lab report. (We strongly suggest that a spreadsheet such as Microsoft® Excel be used to form such tables.)

Parts List

- 100 Ω resistor
- 120 Ω resistor
- 470 Ω resistor
- 1 kΩ resistor
- 1.2 kΩ resistor
- 4.7 kΩ resistor

Procedure

1. Measure all component values.

Note: It is a good habit to always measure the components being used. The results may be in error due to component tolerance errors.

2. Connect the circuit shown in Figure 5-1 and measure the voltage across each resistor. Measure the current into and out of each resistor. Use a VOM for the current measurements and a DMM for the voltage measurements.

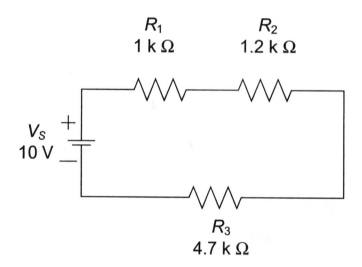

Figure 5-1   Test circuit 1

Table 5-1   Results Table for circuit shown in Figure 5-1

Item	Nominal Value	Predicted Value	Measured Value	Percent Difference
V_S	10 V	10 V		
R_1	1 kΩ	1000 Ω		
R_2	1.2 kΩ	1200 Ω		
R_3	4.7 kΩ	4700 Ω		
R_T				
V_{R1}				
V_{R2}				
V_{R3}				
P_{R1}				
P_{R2}				
P_{R3}				
I_T				

3. Use the VOM for the voltage measurements and the DMM for the current measurements for the circuit shown in Figure 5-2.

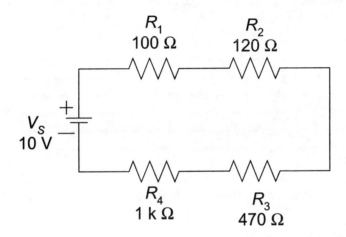

Figure 5-2 Test circuit 2

Table 5-2: Results Table for the circuit shown in Figure 5-2

Item	Nominal Value	Predicted Value	Measured Value	Percent Difference
V_S	10 V	10 V		
R_1	100 Ω	100 Ω		
R_2	120 kΩ	120 Ω		
R_3	470 Ω	470 Ω		
R_4	1 kΩ	1000 Ω		
R_T				
V_{R1}				
V_{R2}				
V_{R3}				
V_{R4}				
P_{R1}				
P_{R2}				
P_{R3}				
P_{R4}				
I_T				

Questions

1. What was the tolerance and range of values for each resistor used? (Analysis)

_____
_____
_____
_____
_____
_____
_____

2. Were the measured values of resistance in tolerance? (Analysis)

_____
_____
_____
_____
_____
_____
_____

3. Were the voltage measurements within an acceptable range compared to the predicted voltages? (Analysis)

_____
_____
_____
_____
_____
_____
_____

4. How did the current in test circuit 2 differ from the current in test circuit 1? (Analysis)

_____
_____
_____
_____
_____
_____
_____

5. What is the smallest power rating that could be used to satisfy the power dissipation needs of all the resistors used? (Analysis)

_____
_____
_____
_____
_____
_____
_____

6. Did the current going into the resistor differ from the current leaving the resistor? What does this tell us? (Conclusion)

_____
_____
_____
_____
_____
_____
_____

7. Which meter was easier to use? Why? (Conclusion)

_____
_____
_____
_____
_____
_____
_____

Experiment 6: Series Circuits, KVL, and Power

Learning Objectives

After completing this laboratory experiment, you should be able to:

- Verify Kirchhoff's Voltage Law (KVL) in a DC series circuit.
- Determine power dissipation in a DC series circuit.

Background

KVL is a special case representation of the Law of Conservation of Energy, which states:

> *Energy can neither be created nor destroyed, but may be transferred and may be converted from one form to another.*

Recall also that electricity is a form of energy. Electrical energy is related to voltage, and voltage is the electrical potential energy difference per unit charge.

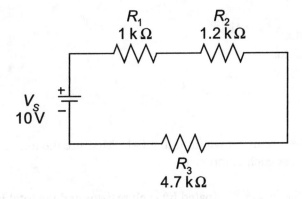

Figure 6-1 Test circuit 1

If one were to start at the positive node of the source and "walk" around the circuit in Figure 6-1, the Law of Conservation of Energy would dictate the following equation:

$$E_S = E_1 + E_2 + E_3$$

Divide all terms by charge:

$$\frac{E_S}{Q} = \frac{E_1}{Q} + \frac{E_2}{Q} + \frac{E_3}{Q}$$

But voltage is the electrical potential energy difference per unit charge:

$$V_s = V_1 + V_2 + V_3$$

Thus KVL states the sum of all voltage rises must equal the sum of all voltage drops.

Prelaboratory Preparation

Predict the total circuit current for each test circuit. Predict the voltage across each resistor in the test circuits. Verify KVL. Predict the power that each resistor will dissipate and the total power dissipated. Record all calculations in the Theoretical Solution section of your engineering notebook. The predicted results should be entered in a results table in the Results section. (You must design the results table.) The results table must also have columns for measured results.

Use an electronic simulation program such as MultiSIM to verify the theoretical calculations.

Parts List

- 1 kΩ resistor
- 1.2 kΩ resistor
- 4.7 kΩ resistor

Procedure

1. Connect the circuit shown in Figure 6-1. Measure the total current. Measure the voltage across each component.

2. Calculate the power dissipated by each resistor and the total power.

3. Verify that the measured results for this circuit satisfied KVL.

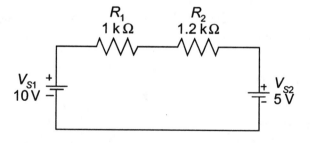

Figure 6-2 Test circuit 2

40

4. Repeat all measurements for the circuits shown in Figures 5-2 and 5-3.

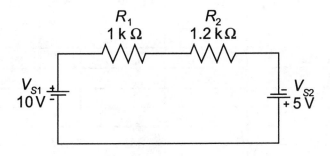

Figure 6-3 Test circuit 3

Questions

1. How many voltage rises and how many voltage drops were in the circuit in Figure 6-2?
2. How many voltage rises and how many voltage drops were in the circuit in Figure 6-3?
3. Comparing questions 1 and 2, are the answers different? Explain why or why not.
4. Was KVL satisfied for each circuit? Show the results.
5. What was the percentage difference for all voltage and current measurements compared to predicted values? Use a spreadsheet to perform these calculations.
6. What is the smallest resistor power rating that could be used in this experiment?

Table 6-1: Data for Circuit 6-2

Component /Measurement	Theoretical Value	Simulated Value	Measured Value	Percentage Different Theoretical to Measured	Percentage Different Simulated to Measured
V_{S1}	+10 V	+10 V			
R_1	1 kΩ	1 kΩ			
R_2	1.2 kΩ	1.2 kΩ			
V_{S2}	-5V	-5V			
V_{R1}					
V_{R2}					
I_T					
P_{R1}					
P_{R2}					

Table 6-2: Data for Test Circuit 6-3

Component /Measurement	Theoretical Value	Simulated Value	Measured Value	Percentage Different Theoretical to Measured	Percentage Different Simulated to Measured
V_{S1}	+10 V	+10 V			
R_1	1 kΩ	1 kΩ			
R_2	1.2 kΩ	1.2 kΩ			
V_{S2}	+5V	+5V			
V_{R1}					
V_{R2}					
I_T					
P_{R1}					
P_{R2}					

Title _____ Page No. ____

Signature _____ Witness _____ Date _____

43

Title _____ Page No. _____

Signature _____ Witness _____ Date _____

Title _____ Page No. _____

Signature _____ Witness _____ Date _____

45

Title _____ Page No. _____

Signature _____ Witness _____ Date _____

Experiment 7: Series Circuits, Voltage Divider, and Meter Loading

Learning Objectives

After completing this laboratory experiment, you should be able to:

- Predict voltages in a voltage divider circuit.
- Describe the effects of meter loading on a circuit.

Background

The voltage divider rule can be thought of as a special case of Ohm's law. The voltage divider circuit is used in circuit applications that require lower voltages than that supplied by the source.

The equation for the voltage divider rule is:

$$V_X = R_X \frac{V_T}{R_T}$$

Note that V_T/R_T is the total current. Therefore, the voltage divider rule is a convenient form of Ohm's law, which allows for faster calculations and provides more insight into series circuits.

The idea that measurement equipment affects the measurement can be difficult to grasp. Although the equipment we use to make measurements should have a minimal effect on the item being observed, this may not always be the case. All measuring equipment has an internal resistance, which is seen at the input terminals of the equipment. A DMM may have an input resistance of 10 MΩ. An analog meter has an input resistance rating that is specified in Ω/V. This quantity is shown on the meter face. It may be 100,000 Ω/VDC, which means that on the 2.5 V scale, the meter has 250,000 Ω of resistance. This is the resistance that will be in parallel with the component being measured.

Prelaboratory Preparation

For each test circuit, predict the voltage across each resistor using the voltage divider rule. Verify KVL for each circuit. Generate a results table to record the predictions and the measured results. Also use a circuit simulation program to verify the predicted results.

Parts List

- 1 kΩ resistor
- 1.2 kΩ resistor

- 2 kΩ resistor
- 2.2 kΩ resistor
- 3.3 kΩ resistor
- 10 kΩ resistor
- 22 kΩ resistor
- 33 kΩ resistor
- 10 kΩ potentiometer
- (Qty. 2) 1 MΩ resistors

Procedure

1. Connect the circuit shown in Figure 7-1. Measure the voltage across R_1 and R_2. Compare the measurements to what the values should be using nominal values, based on the voltage divider rule. Explain any differences.

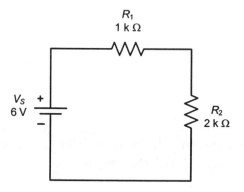

Figure 7-1 Test circuit 1

2. Repeat Step 1 for all resistors in the circuits shown in Figures 7-2 and 7-3.

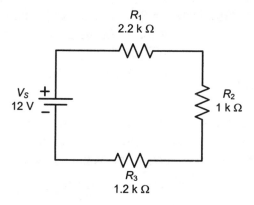

Figure 7-2 Test circuit 2

48

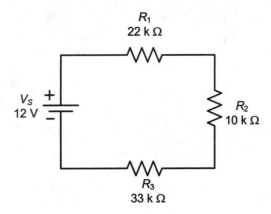

Figure 7-3 Test circuit 3

3. Measure the voltages across R_1 and R_2 in Figure 7-4. Start with R_2 at a minimum, and then turn the shaft to ¼ of its full range, then ½, then ¾. Compare the results to the voltage divider rule prediction. Explain any differences. Determine what R_2 will need to be set to for a voltage drop of 5 V. Verify your prediction with a measurement.

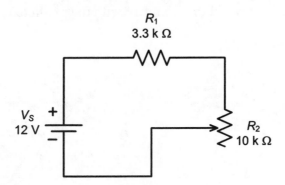

Figure 7-4 Test circuit 4

4. Connect the circuit in Figure 7-5. Use a DMM and measure and record the voltage across R_2. Now use a VOM and measure and record the voltage across R_2. Use the most sensitive scale possible. Comment on any differences from the predicted value in both cases.

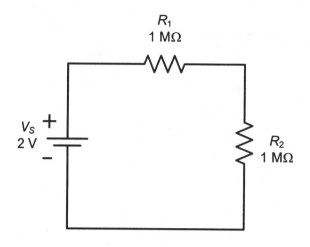

Figure 7-5 Test circuit 5

Questions

1. Compare predicted to measured values for all data taken. Explain any differences.
2. What is the purpose of a voltage divider circuit? How can a variable voltage divider circuit be constructed?
3. Comment on the internal resistance of a DMM and a VOM based on your measurement results.
4. What was the voltage across R_2 for the circuit shown in Figure 7-5 when the VOM was used? How did this compare with the predicted results? Explain.

Title _____ Page No. ___

Signature _____ Witness _____ Date _____

Title _____ Page No. _____

Signature _____ Witness _____ Date _____

Title _____ Page No. ___

Signature _____ Witness _____ Date _____

53

Title _____ Page No. _____

Signature _____ Witness _____ Date _____

Experiment 8: Parallel Circuits and Ohm's Law

Learning Objectives

After completing this laboratory experiment, you should be able to:

- Predict and measure the total resistance in parallel circuits.
- Demonstrate Ohm's law as applied to parallel circuit.
- Predict and measure voltages and currents in parallel circuits.

Background

Recall that in a series circuit there is only one path for current flow. The current is the same throughout the circuit. The source voltage is divided among the elements. In a parallel circuit, the current has multiple paths. Examine the circuit in Figure 8-1.

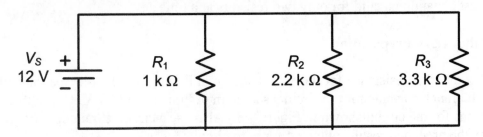

Figure 8-1   A parallel circuit

If one were to trace the current from the positive terminal of the source, through the resistors, and back to the negative terminal, there would be three current paths. The voltage measured across each resistor would equal the source voltage.

The total current would be the sum of all the branch currents. The branch currents are the currents through each resistor.

$$I_T = I_1 + I_2 + I_3$$

Using Ohm's law, the current equation can be rewritten as:

$$\frac{V_T}{R_T} = \frac{V_1}{R_1} + \frac{V_2}{R_2} + \frac{V_3}{R_3}$$

Since the voltage drops are equal, $V_S = V_1 = V_2 = V_3$, the above equation can be modified to:

$$\frac{1}{R_T} = \frac{1}{R_1} + \frac{1}{R_2} + \frac{1}{R_3}$$

This yields an equation to calculate the total resistance:

$$R_T = \frac{1}{\dfrac{1}{R_1} + \dfrac{1}{R_2} + \dfrac{1}{R_3}}$$

If there are only two resistors in parallel (or one can group parallel resistors two at a time) then the following equation can be used:

$$R_T = \frac{R_1 R_2}{R_1 + R_2}$$

The last equation is valid for only two resistors at a time.

Prelaboratory Preparation

Predict the total resistance for the circuit shown in Figure 8-2. Predict the total current and all branch currents for the circuits shown in Figures 8-1 and 8-3. Use both switch positions for the circuit shown in Figure 8-3. Also use a circuit simulation program to verify the predicted results obtained for both circuits.

Parts List

- 820 Ω resistor
- 1 kΩ resistor
- 2.2kΩ resistor
- 3.3 kΩ resistor
- 6.8 kΩ resistor
- 10 kΩ resistor

Procedure

1. Measure the total resistance of the circuit shown in Figure 8-2 and compare to the predicted value.

2. Add the supply as shown in Figure 8-1 and measure all currents and the voltage.

3. Add a 10 kΩ resistor as R_4 in parallel. Repeat all measurements.

4. Replace the 10 kΩ resistor with an 820 Ω resistor and repeat all the measurements.

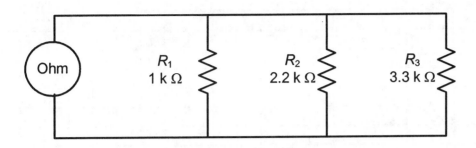

Figure 8-2 Test circuit 1

5. Connect the circuit in Figure 8-3. With the switch open, measure the total resistance, total current, and current through each resistor. Close the switch and repeat the process.

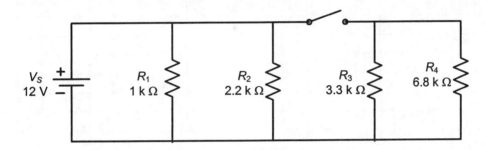

Figure 8-3 Test circuit 2

Questions

1. Verify Ohm's law using measured values for each resistor for the circuits tested.
2. What happened when the 10 kΩ resistor was added in parallel to the circuit shown in Figure 8-1? Explain why.
3. What happened when the 10 kΩ resistor was changed to an 820 Ω resistor? Explain why.
4. In the last circuit, explain what happened when the switch was closed. What were the effects seen on the circuit?

Title _____ Page No. ____

Signature _____ Witness _____ Date ____

58

Title _____ Page No. __

Signature _____ Witness _____ Date _____

59

Title _____ Page No. ____

Signature _____ Witness _____ Date _____

60

Title _____ Page No. ___

Signature _____ Witness _____ Date _____

Title _____ Page No. ____

Signature _____ Witness _____ Date _____

62

Experiment 9: Parallel Circuits, KCL and Current Divider

Learning Objective

After completing this laboratory experiment, you should be able to:

- Apply the current divider rule and verify Kirchhoff's current law in parallel circuits.

Background

Recall that a parallel circuit has multiple current paths. Therefore the current will split between the paths. How does this split occur? Examine the circuit in Figure 9-1.

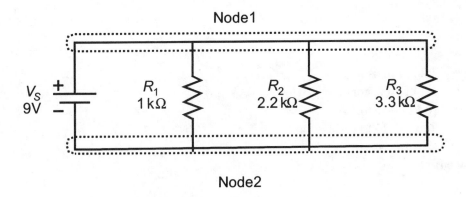

Figure 9-1 Parallel test circuit

There are two "nodes" in this circuit, that is, two points where all the components connect. We can examine the currents at both nodes. How many currents are entering node 1 and how many are leaving node 1? There is one current entering node 1, coming from the 9 V source. There are three currents leaving node 1, going to the resistors. The opposite is true at node 2. KCL states that the sum of the currents entering a node must equal the sum of currents leaving a node. In Figure 9-1, the total current enters node 1. The currents to all resistors, that is, the branch currents, must add up to the total current.

This division of current leads one to think of a current divider rule. The current is divided in an inverse relationship to the branch resistance. The higher the branch resistance, the lower the current. One only needs to know the total current, total resistance, and branch resistance to find the branch current.

The current divider rule is:

$$I_X = \frac{I_T R_T}{R_X}$$

Note that $I_T R_T$ is the voltage across the parallel circuit. When this voltage is divided by R_X, which forms Ohm's law, then the current in the branch is known. The current divider rule is useful when the total current is known and the voltage is unknown. One step is saved.

Prelaboratory Preparation

Predict the voltage, branch currents, and total current for each test circuit. Generate a results table to record the predicted values and the measured values. Also use a circuit simulation program to verify the predicted results.

Parts List

- 1 kΩ resistor
- 2.2 kΩ resistor
- 3.3 kΩ resistor
- 10 kΩ resistor
- (Qty. 2) 20 kΩ resistors

Procedure

1. Connect the circuit in Figure 9-1. Measure all voltages and currents in all branches. Compare the measured values with the predicted values in your engineering notebook and verify KCL.

2. Connect the circuit in Figure 9-2. Measure the total current and both branch currents. Use this measured data to verify KCL and the current divider rule.

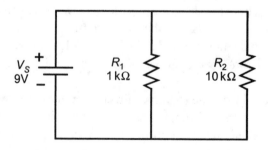

Figure 9-2   Current divider test circuit

3. Connect the circuit in Figure 9-3. Measure all branch currents. Use the
    measured data to verify KCL for this circuit.

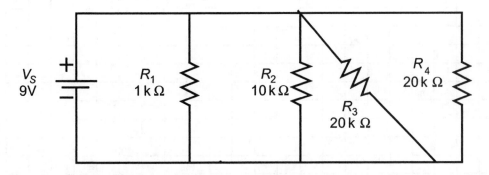

Figure 9-3 KCL test circuit

Questions

1. What was the percentage difference of the measured resistor values when
    compared to the nominal resistor values?
2. Was Kirchhoff's current law verified for all circuits?
3. Was the current divider rule verified?

Title _____ Page No. __

Signature _____ Witness _____ Date _____

66

Title _____ Page No. __

Signature _____ Witness _____ Date _____

Title _____ Page No. __

Signature _____ Witness _____ Date _____

Title _____ Page No. ___

Signature _____ Witness _____ Date _____

Title _____ Page No. __

Signature _____ Witness _____ Date _____

70

Experiment 10: Series–Parallel Circuits

Learning Objective

After completing this laboratory experiment, you should be able to:

- Predict and measure voltages and currents in series–parallel DC circuits.

Background

You have examined series circuits and parallel circuits separately in previous experiments. These are not the only circuits in electronics. When one measures the voltage across a resistor, a parallel circuit is formed with the voltmeter. When one measures the current in a branch, a series circuit is formed with the ammeter. In this experiment the circuit is set up to have both series and parallel elements. The general strategy to solve such a circuit is:

- Identify groups of components in parallel and groups of components in series.
- Apply series–parallel circuit analysis techniques appropriately to series and parallel groups.

As an example, consider the circuit in Figure 10-1. In this circuit R_1 and R_2 are in parallel. That combination is in series with R_3. Reducing the parallel combination would yield an R_{EQ} of 500 Ω. The total resistance would then be 1.5 kΩ. One could find the total current and all voltages. The branch currents could also be found.

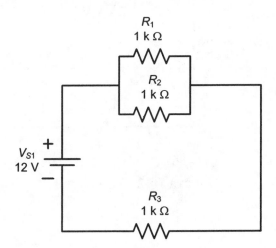

Figure 10-1   Test circuit 1

Prelaboratory Preparation

Calculate all voltages and currents for each test circuit. Generate a results table to record the predicted values and the measured values. Also use a circuit simulation program to verify the predicted results.

Parts List

- (Qty. 3) 1 kΩ resistors
- 1.2 kΩ resistor
- 10 kΩ resistor
- 22 kΩ resistor

Procedure

1. Connect the circuit shown in Figure 10-1 and verify the total resistance. Measure all voltages and currents. Compare these to the predicted values.

2. Connect the circuit shown in Figure 10-2 and measure all voltages, currents, and resistances of all resistor combinations. Compare these results to the predicted results. Be sure to redraw the circuit in your engineering notebook to show the equivalent resistances.

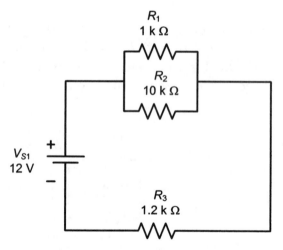

Figure 10-2 Test circuit 2

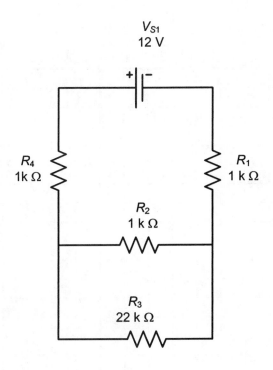

Figure 10-3 Test circuit 3

3. Repeat the process for the circuit shown in Figure 10-3.

Question

1. What is the percentage difference between the measured and predicted values of total resistance?
2. If an analog meter were used to measure voltages in the circuit shown in Figure 10-3, would there be a meter loading effect? Explain.
3. Use KVL to verify the voltage drops in the circuit shown in Figure 10-2.
4. Use KCL to verify the branch currents in the circuit shown in Figure 10-3.
5. In general, how is the analysis of a series–parallel circuit performed?

Title _____   Page No. ___

Signature _____   Witness _____   Date _____

Title _____ Page No. ___

Signature _____ Witness _____ Date _____

Title _____ Page No. _____

Signature _____ Witness _____ Date _____

76

Title _____ Page No. __

Signature _____ Witness _____ Date _____

Title _____ Page No. _____

Signature _____ Witness _____ Date _____

Experiment 11: Superposition in a Multiple-Source DC Circuit

Learning Objective

After completing this laboratory experiment, you should be able to:

- Apply superposition in multiple-source DC series–parallel circuits.

Background

A circuit may have several signal sources. This experiment will concentrate on multiple-source DC circuits as shown in Figure 11-1.

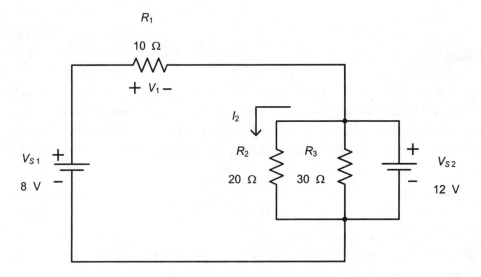

Figure 11-1 Test circuit 1

The general strategy to use superposition in the analysis of multiple-source circuits is:

- Label the desired quantities to be found.
    - Assign an arbitrary polarity to each voltage to be found.
    - Assign an arbitrary direction to each current to be found.
- Deactivate all sources expect for one.
    - Voltage sources are replaced by a short circuit.
    - Current sources are replaced by an open circuit.
- Analyze the circuit using series–parallel analysis techniques. Be careful with polarities and directions.
- Deactivate the first source and reactivate another source. Perform the circuit analysis again.
- Repeat the process for each remaining source.

- Add the corresponding resultant voltages and currents from the analysis with each source. Be sure to add the results with the proper signs.

Prelaboratory Preparations

Predict the current through R_1, R_2, and R_3 for the circuits in Figures 11-2 and 11-3 using superposition. Calculate the voltages across R_1, R_2 and R_3 for each circuit using superposition. Also use a circuit simulation program to verify the predicted results. Generate a results table to record both the predicted and measured values.

Parts List

- 6.8 kΩ resistor
- 8.2 kΩ resistor
- 22 kΩ resistor

Procedure

Connect test circuits 2 and 3 and measure the current through each resistor with both sources in the circuit. Then measure the voltage across each resistor. Repeat these measurements with V_{S2} deactivated. Repeat again with V_{S1} deactivated. Use superposition to verify the voltages and currents measured with both sources in the circuit.

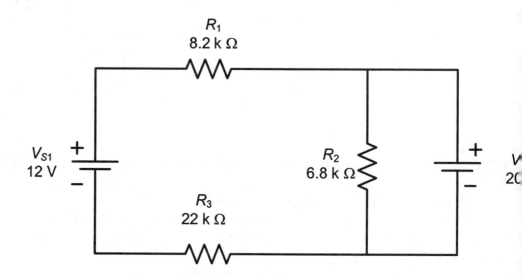

Figure 11-2   Test circuit 2

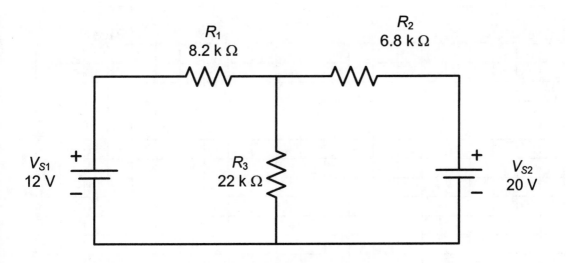

Figure 11-3   Test circuit 3

Questions

1. Compare all predicted values with measured values.
2. What is the percentage difference? Is this percentage difference acceptable?
   What caused the percentage difference?
3. Compare the circuits in Figures 11-2 and 11-3. Both circuits have the same
   voltage sources. Why are the resistor voltages and currents different?
4. Was superposition verified for a multiple-source DC circuit? Explain.

Title _____ Page No. ___

Signature _____ Witness _____ Date _____

Title _____ | Page No. _____

Signature _____ Witness _____ Date _____

Title _____ Page No. ____

Signature _____ Witness _____ Date ____

Title _____ Page No. ___

Signature _____ Witness _____ Date _____

Title _____ Page No. ____

Signature _____ Witness _____ Date _____

Experiment 12: AC Signals

Learning Objectives

After completing this laboratory experiment, you should be able to:

- Describe an AC sinusoidal signal.
- Determine peak and RMS amplitudes, and radian and cyclic frequencies.
- Operate an oscilloscope as a measuring instrument.
- Operate a function generator.

Background

The previous experiments have concentrated on DC sources. Electricity is not DC only. There are many AC signal sources. The music we listen to is a form of AC. The electricity from the wall outlet is AC. Radio and TV signals are AC. Knowing there are various types of AC signals, we can benefit from learning about AC signals and how to measure them.

There is one basic AC signal to be most concerned with, the sine wave as shown in Figure 11-1. The sine wave is expressed by the following equation.

$$v(t) = V_p \sin(\omega t)$$

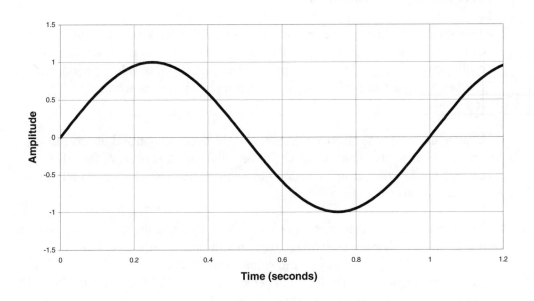

Figure 12-1 Graphical representation of the sine wave

87

A function generator can generate a sine wave. The sine wave repeats every 360° or every 2π radians. The radian frequency is ω in rad/s and is equal to $2\pi f$, where f is the cyclic frequency in Hz.

The function generator that will be used in this experiment can also generate square waves and triangle waves. You will learn in later courses that these two waves are just a combination of sine waves of different frequencies. Hence, learning the sine wave is the key to AC.

Prelaboratory Preparation

Read the operator's manual for the oscilloscope and the function generator. Summarize key ideas and operational notes in your engineering notebook (one page maximum). (How is the function generator properly connected to the oscilloscope?)

Procedure

1. Turn the instrument on. Set the oscilloscope for single trace operation. Set the triggering to internal trigger and the time base to 0.1 ms/div. With the input selector set to GND, observe a single horizontal line on the display. Rotate the vertical position control back and forth and observe the position of the line. Adjust the brightness and focus controls while observing the effects. *Caution: An overly bright trace may damage the screen permanently, particularly if the trace is a single nonmoving dot.*

2. Turn on the function generator and set it to 1 kHz with a sine wave output. Connect the function generator to the oscilloscope. Set the input selector of the oscilloscope to AC. Measure the peak value, peak-to-peak value, period, and frequency (if possible) of the signal. Determine both the radian and cyclic frequencies.

3. Change the time base (seconds/division) of the oscilloscope and observe the effects. Change the input sensitivity (Volts/division) and observe the effects. Change the function generator output to a square wave and then a triangle wave. Observe the effects.

4. Adjust the output of the function generator to 1 V_{P-P} sine wave as measured on the oscilloscope. Measure the voltage using a voltmeter. Adjust the function generator to 2 V_{P-P} and measure the voltage on a voltmeter. Repeat for 5 V_{P-P}. Adjust the frequency of the function generator to 10 kHz, 100 kHz, and 200 kHz, and repeat all voltage measurements. Record the results.

5. Disconnect the function generator. Now set a DC supply to 5 V and connect the positive terminal to the oscilloscope input and the negative terminal to the

oscilloscope ground. Set the input sensitivity to 5 V/div. Leave the input selector at AC. Record the result. Set the input selector to DC. Record the result.

6. **If your instructor approves of this step, taking into consideration your particular power supply**, connect the negative terminal of the power supply to the oscilloscope input and the positive terminal of the power supply to the oscilloscope ground. Record the results. Adjust the supply voltage and observe the results.

Questions

1. Explain how to determine peak and peak-to-peak voltages from the oscilloscope display.
2. Explain how to determine frequency from the oscilloscope trace.
3. What was the result of measuring DC with the input selector set to AC? Why?
4. Did the AC reading from the oscilloscope differ from the voltmeter? Why? Reconcile the differences mathematically.
5. What are some points of safety listed in the operator's manuals?

Title _____ Page No. __

Signature _____ Witness _____ Date _____

Title _____ Page No. __

Signature _____ Witness _____ Date _____

Title _____ Page No. ___

Signature _____ Witness _____ Date _____

92

Experiment 13: Capacitors in DC and AC Circuits

Learning Objectives

After completing this laboratory experiment, you should be able to:

- Describe and predict the fundamental action of capacitors in DC circuits.
- Describe and predict the fundamental action of capacitors in AC circuits.

Background

DC Transient

A capacitor is a component that has the property of storing electrical energy in an electric field. Capacitance is measured in farads (F), which is coulombs per volt. A capacitor opposes an instantaneous change in voltage. Why? It takes time to move the charge onto or off the plates of the capacitor. This time is determined from the time constant, τ.

$$\tau = RC$$

where τ has the unit of time, seconds. The equation below is used to determine the voltage across the capacitor at any point in time during charging.

$$v_C(t) = V_f(1 - e^{-t/\tau})$$

V_f is the final value of the voltage across the capacitor. For a charging condition the initial voltage is usually zero. One can see from the equation that v_C becomes constant or close to constant when $t = 5\tau$.

For a discharging capacitor:

$$v_C(t) = V_o e^{-t/\tau}$$

V_o is the initial value of the voltage across the capacitor. The capacitor is fully discharged at five time constants. The test circuit in Figure 13-1 can be used to demonstrate the charging and discharging characteristics as predicted by the equation for $v_C(t)$ and shown in Figure 13-2. This is a general curve. Scale the y-axis values by either V_f or V_o.

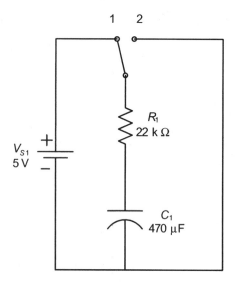

Figure 13-1 Test circuit

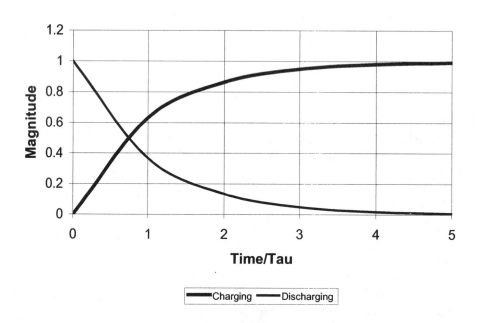

Figure 13-2 Charging and discharging curves for a capacitor
(Tau = τ)

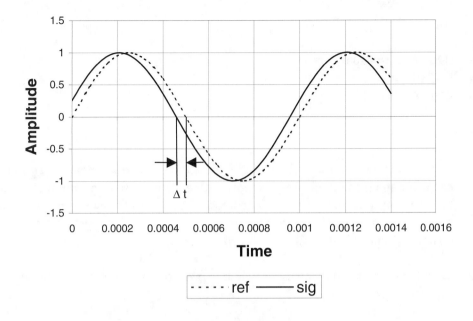

Figure 13-3   Phase shift between the two AC signals

<u>AC Steady-State</u>

Capacitors can cause phase shift in an AC circuit. To determine the phase shift caused by the RC circuit, measure Δt as shown in Figure 13-3. Then measure the period T, which is the time for one complete cycle. The phase shift in degrees is then found from:

$$\text{Phase Shift} = \theta = 360° \frac{\Delta t}{T}$$

Prelaboratory Preparation

Predict the charging and discharging curves for the capacitor voltage in the circuit shown in Figure 13-1.   Create separate curves for each resistor value listed in the procedure.

Predict the voltages for each component in Figure 13-4. Explain how the voltages were determined. *Hint*: Use Ohm's law and the steady-state operation of the capacitor.

Predict the voltages with phase shift for the circuit in Figure 13-6. Use all the frequencies listed in the procedure.

Parts List

- (Qty. 3) 1 kΩ resistors
- 10 kΩ resistor
- 22 kΩ resistor

- (Qty. 3) 1 µF capacitors
- 0.47 µF capacitor
- 470 µF capacitor

Procedure

1. Connect the test circuit as shown in Figure 13-1. Place a voltmeter across the capacitor and turn on the power supply with the switch set in position 2. Ensure that the supply voltage is 5 V. Keep track of time once the switch is moved to position 1. Measure the voltage at various time intervals to reproduce the charging graph. (What time interval should be used?)

2. With the capacitor fully charged switch to position 2 and repeat the measurement process.

3. Change the 22 kΩ resistor to 10 kΩ and repeat the process.

4. Now connect Figure 13-4. Measure the voltages across all the resistors and all the capacitors. You may need to turn power off and discharge all capacitors between measurements, especially if you use a VOM. A DMM is recommended for these measurements, if available.

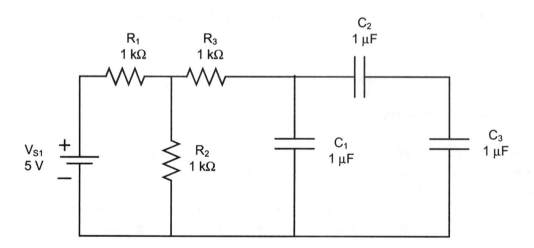

Figure 13-4 Capacitor test circuit

5. Connect the test circuit in Figure 13-5. First measure the voltage waveform across the resistor on channel 2 and the voltage across the source on channel 1. The source should be set to 1 V_{P-P}, 100 Hz, and a square wave. Sketch the display of the oscilloscope. Then measure the voltage waveform across the capacitor.

Note: The position of the capacitor and the resistor need to be switched to make this measurement.

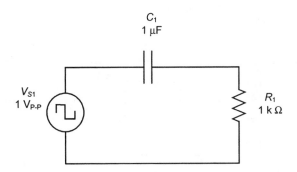

Figure 13-5   Charge, discharge test circuit

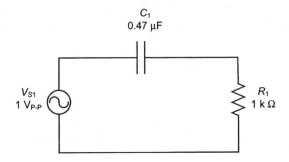

Figure 13-6   Phase-shift test circuit

6. Now switch the signal generator to a sine wave, as shown in Figure 13-6 and set the frequency to 20 Hz. Measure the phase shift. Repeat the phase-shift measurement at frequencies of 200 Hz, 500 Hz, 1 kHz, and 2 kHz. Switch the position of the resistor and the capacitor and repeat the measurements.

Questions

1. Compare the actual graphs generated with the test circuit in Figure 13-1 to the theoretical graph. Explain any differences.
2. Compare the predicted values to the measured values and explain any differences in Figure 13-4.
3. What is the phase-shift trend versus frequency? Explain.
4. Explain the waveforms obtained when the square wave was applied to the RC circuit.
5. Judging from your instrumentation specifications, can you measure the differences between $v_C(5\tau)$ and V_f in a charging RC circuit?

Title _____ Page No. __

Signature _____ Witness _____ Date _____

Title _____   Page No. __

Signature _____   Witness _____   Date _____

Title _____ Page No. __

Signature _____ Witness _____ Date _____

Title _____ Page No. __

Signature _____ Witness _____ Date _____

Title _____ Page No. __

Signature _____ Witness _____ Date _____

Experiment 14: Inductors in DC and AC Circuits

Learning Objectives

After completing this laboratory experiment, you should be able to:

- Describe and predict the fundamental action of inductors in DC circuits.
- Describe and predict the fundamental action of inductors in AC circuits.

Background

DC Transient

The previous experiment covered an energy storage device, the capacitor. This experiment introduces another energy storage device, the inductor. You will examine the DC and the AC properties of the inductor. The inductor is an energy storage device that stores electrical energy in a magnetic field. The inductor opposes instantaneous changes of current. (Why?)

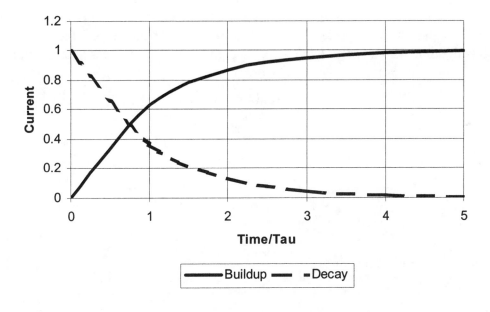

Figure 14-1 Buildup and decay curves for an inductor

There are buildup and decay plots just as there were for the capacitor, as shown in Figure 13-1. The difference is in the component that stores energy. The inductor stores energy in a magnetic field and the magnetic field is proportional to current. The capacitor stores energy in an electric field and the electric field is proportional to voltage.

The transient current equation during buildup is:

$$i(t) = I_F \left(1 - e^{-t/\tau}\right)$$

where τ is L/R (units: s) and I_F is the final value of the current through the inductor.

A note of caution: The primary failure modes of an inductor are shorts between the windings and open circuits in the windings due to excessive currents, overheating, and aging. Opens are easy to identify with an ohmmeter. A short is more difficult to identify due to the low DC resistance of most inductors. An LCR meter is needed to test for the inductance value.

The equation for decay is:

$$i(t) = I_i e^{-t/\tau}$$

Again τ is L/R. I_i is the initial value of the current through the inductor. The buildup and decay have reached a steady-state value at $t = 5\tau$.

AC Steady-State

Inductors can cause a phase shift in an AC circuit. To determine the phase shift caused by the RL circuit, measure Δt as shown in Figure 13-2. Then measure the period T, which is the time for one complete cycle. The phase shift in degrees is then found from:

$$\text{Phase Shift} = \theta = 360^\circ \frac{\Delta t}{T}$$

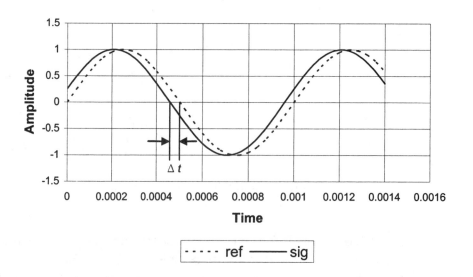

Figure 14-2 Phase shift between two AC signals

Prelaboratory Preparation

Predict the charging and discharging curves for the current through and the voltage across the inductor for the test circuit in Figure 14-3.

For an AC series circuit consisting of a 100 mH coil and 1 kΩ resistor, predict the voltage drop across the resistor (including phase shift) for frequencies of 500 Hz, 1 kHz, 2 kHz, and 4 kHz.

Change the inductor to 47 mH, and then predict the voltage drop across the resistor (including phase shift) for frequencies of 2 kHz, 5 kHz, and 10 kHz.

Parts List

- 1 kΩ resistor
- 100 mH inductor
- 47 mH inductor

Procedure

1. Connect the test circuit shown in Figure 14-3 and set the source frequency to 500 Hz. Measure and sketch the voltage across R_1. Calculate the current through the inductor and the voltage across the inductor. Set the frequency to 1 kHz and repeat the measurement, and the voltage and current calculation. Repeat the measurements at frequencies of 2 kHz and 4 kHz.

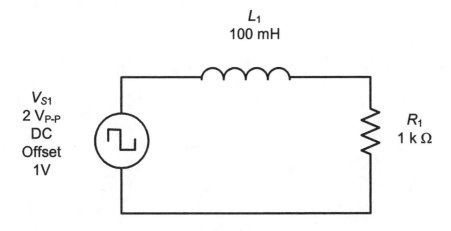

Figure 14-3 Test circuit

2. Change the generator to produce a sine wave and measure the phase shift at the three frequencies used in step 1.

3. Change L_1 to 47 mH. Repeat the phase measurements for frequencies of 2 kHz, 5 kHz, and 10 kHz.

Questions

1. Calculate the reactance of the inductor from the voltage and current measurements. How do these compare to the theoretical reactance values? Explain any differences.
2. Explain the waveforms obtained with the square wave applied to the RL circuit.
3. What is the phase shift trend versus frequency? What is the phase shift trend versus inductance? Explain both answers.
4. Why was the square wave used to illustrate the transient behavior?
5. Compare the theoretical and experimental transient plots. Explain any significant differences.

Title _____ Page No. ___

Signature _____ Witness _____ Date _____

107

Title _____ Page No. _____

Signature _____ Witness _____ Date _____

Title _____ Page No. ____

Signature _____ Witness _____ Date _____

Title _____ Page No. ___

Signature _____ Witness _____ Date _____

Experiment 15: KVL and Voltage Divider in AC Series Circuits

Learning Objectives

After completing this laboratory experiment, you should be able to:

- Describe the fundamental properties of series AC circuits.
- Calculate all voltages in single-source AC series circuits.

Background

In Experiments 6 and 7, the topic of KVL and the voltage divider rule were introduced with a DC circuit. This experiment expands those topics to an AC circuit.

KVL states that the sum of all voltage rises equals the sum of all voltage drops around a closed loop. This is regardless of whether it is an AC or DC input. The oscilloscope or DMM can be used in pure resistive AC circuits. In an AC circuit with reactive components the oscilloscope must be used to obtain both the magnitude and the phase of the voltage.

The ground connection to the oscilloscope requires special attention. How does one make voltage drop measurements with an oscilloscope?

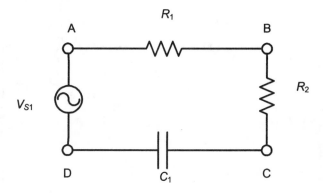

Figure 15-1 AC Series Circuit

If the voltage across R_1 is desired, then the oscilloscope leads are placed across points A and B. Correct? Wrong! Point B would then be at ground and R_2 and C_1 would be effectively eliminated (shorted) from the circuit. So how can the voltage drop across R_1 be made? Place R_1 between points C and D and move C_1 to the former position of R_1. Then the scope ground will be at the same point as the generator ground.

Prelaboratory Preparation

Use the voltage divider rule to predict the voltage drops across all components in the circuit shown in Figure 15-2 and Figure 15-3, at all the frequencies specified in the procedure. Create a results table to record the predicted results and measured results. Use KVL to verify the predicted results. Also use a circuit simulation program to verify the predicted results.

Parts List

- 820 Ω resistor
- 1 kΩ resistor
- 2.2 kΩ resistor
- 0.01 μF capacitor

Procedure

1. Connect the circuit shown in Figure 15-2. Set the frequency of the function generator to 200 Hz and measure the voltage across all the components. Use both the oscilloscope and the DMM. Also measure and record the phase shift between the source and each resistor voltage. Repeat at 1 kHz, 2 kHz, and 3 kHz.

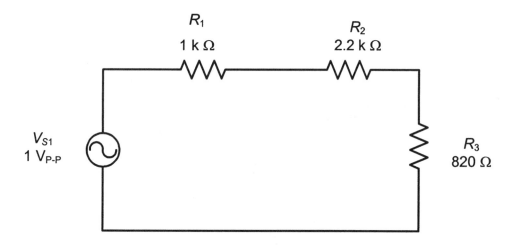

Figure 15-2 Test circuit 1

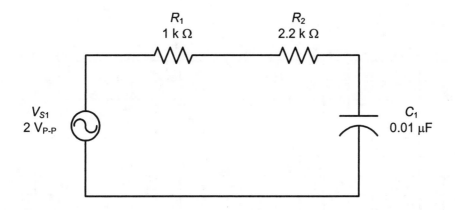

Figure 15-3   Test circuit 2

2. Now connect the circuit shown in Figure 15-3. Measure the voltage across the capacitor and the voltage across all resistors at 2 kHz using both the DMM and the oscilloscope. Measure and record the phase shift between the source and each component. Repeat for 3 kHz.

Questions

1. Does KVL verify the oscilloscope measurements, both magnitude and phase for the first test circuit? Show your results for each frequency.
2. Does KVL verify the measured results for test circuit 1 using only the DMM? Explain.
3. Does KVL verify the measured results for test circuit 2 using only the DMM? Explain.
4. Compare the phase shifts between the corresponding voltages in test circuit 1 and test circuit 2. Explain why there are significant differences.

Title _____ Page No. __

Signature _____ Witness _____ Date ____

Title _____ Page No.

Signature _____ Witness _____ Date _____

115

Title _____ Page No. ___

Signature _____ Witness _____ Date _____

Title _____ Page No. _____

Signature _____ Witness _____ Date _____

Title _____ Page No. ___

Signature _____ Witness _____ Date _____

118

Experiment 16: KCL and Current Divider in AC Parallel Circuits

Learning Objectives

After completing this laboratory experiment, you should be able to:

- Describe the fundamental properties of AC parallel circuits.
- Predict all voltages and currents in single-source AC parallel circuits.

Background

The DC parallel circuit was examined in previous experiments. The AC parallel circuit follows the same rules as the DC parallel circuit.

Recall that a parallel circuit has multiple current paths. Therefore the current will be split between the paths. Examine the parallel circuit shown in Figure 16-1.

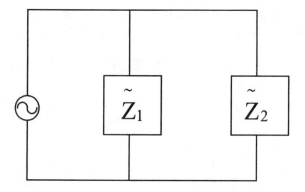

Figure 16-1   The parallel circuit

The impedances of the components, $\tilde{Z}_1$ and $\tilde{Z}_2$, are measured in ohms. The current will divide among the parallel elements. The voltage across the parallel elements will be the same. Knowing voltage and impedance, one can calculate current.

Prelaboratory Preparation

Predict the voltages across and currents through the parallel resistors in the circuits shown in Figure 16-2 and Figure 16-3. Assume $\tilde{V}_t = 1\angle 0°$. Use the current divider rule to predict the currents in R_2 and R_3. Verify these results with KCL. Generate a results table to record the predicted and the measured results. Also use a circuit simulation program to verify the predicted results.

Parts List

- (Qty. 3) 10 Ω resistors
- 820 Ω resistor
- 5.6 kΩ resistor
- 1 kΩ resistor

Procedure

1. Connect the test circuit shown in Figure 16-2. The 1 kΩ resistor is to be considered part of the source for setting the terminal voltage. Set the signal generator to 100 Hz and measure the terminal voltage (after R_1) and the voltage across the parallel components using both the oscilloscope and the DMM. Calculate the branch currents from the voltage measurement. Now measure the branch currents. Then measure the total current. Calculate the total current.

2. Repeat step 1 at 500 Hz and 1 kHz. Verify that V_t has remained constant at each frequency. Adjust V_{S1} if necessary.

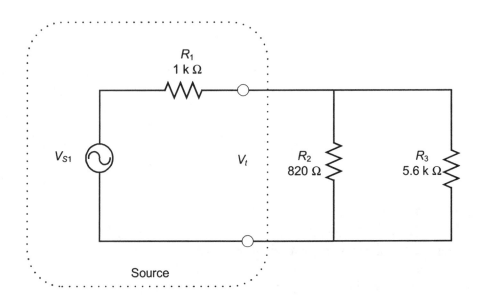

Figure 16-2 Test circuit

120

3. Connect the test circuit shown in Figure 16-3. The 1 kΩ resistor is again to be considered part of the source for setting the output voltage. Set the signal generator to 100 Hz and set the output voltage V_t to 1 V$_{P-P}$. Measure the voltage, **magnitude** and **phase**, across the parallel components with respect to V_{S1}. Then measure the branch currents. It may be helpful to place a 10 Ω sense resistor in series with the C_1 and R_2. Then measure the voltage across the sense resistor and calculate the branch current. The same process can be used to measure the total current. Place the sense resistor in the ground return path.

4. Repeat step 1 at 500 Hz and 1 kHz.

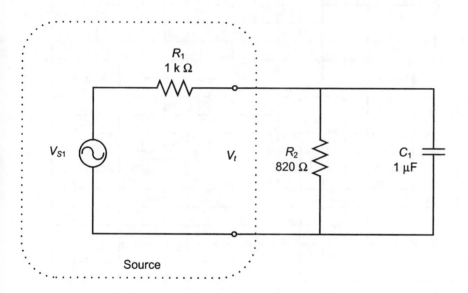

Figure 16-3 Test circuit

Questions

1. Verify the current divider rule for the three frequencies using the measured results, in both test circuits.
2. Verify KCL for the three frequencies using the measured results, in both test circuits.
3. What were the phase shifts between the resistor voltages and the source voltage (V_t) in the circuit shown in Figure 16-2? Explain.
4. What were the phase shifts between the resistor voltage (R_2) and the source voltage (V_t) in the circuit shown in Figure 16-3? Explain
5. Could the generator be damaged if R_1 in Figures 16-2 and 16-3 is removed from the circuit? How?

Title _____ Page No. __

Signature _____ Witness _____ Date _____

Title _____ Page No. __

Signature _____ Witness _____ Date _____

Title _____ Page No. __

Signature _____ Witness _____ Date _____

124

Title _____ Page No. __

Signature _____ Witness _____ Date _____

125

Title _____ Page No. __

Signature _____ Witness _____ Date _____

Experiment 17: AC Series Circuits with L and C

Learning Objectives

After completing this laboratory experiment, you should be able to:

- Use phasors to determine AC voltages, currents, and impedances of components and the entire circuit in series circuits with reactance.
- Describe magnitude and phase relationships for voltages, currents, and impedances in series circuits with reactance.

Background

The inductor and capacitor act as energy storage devices, even at AC. Both components have reactance, limit current flow, and have frequency dependent characteristics. For example, the reactance of the inductor varies with frequency as:

$$X_L = 2\pi f L$$

And the reactance of the capacitor varies with frequency as:

$$X_C = \frac{1}{2\pi f C}$$

In a series circuit the impedance of the components adds, which is similar to resistors in series. Ohm's law will hold true. *Caution*: When dealing with reactances, pay special attention to the phase angles of the quantities involved. Remember that these impedances carry a $\pm 90^\circ$ phase angle.

Something special happens at the point where $X_L = X_C$ in a series circuit. If $X_L = X_C$, the impedances of the inductor and the capacitor cancel each other. (Why?) The resulting total impedance is then real. The resulting total impedance has 0° of phase shift. This concept will also be expanded on in later experiments.

Prelaboratory Preparation

Predict the voltage, both magnitude and phase, across each component in Figure 17-1 for all frequencies indicated in the procedure. Repeat for $C = 0.47$ μF. Also use a circuit simulation program to verify the predicted results.

Parts List

- 1 kΩ resistor
- 0.1 µF capacitor
- 0.47 µF capacitor
- 47 mH inductor

Procedure

1. Connect the circuit shown in Figure 17-1. Set the frequency of the function generator to 1 kHz and measure the voltage across each component. Include any phase information.

2. Then adjust the frequency in 200 Hz increments and record the voltage across the resistor. Continue until 3 kHz has been reached.

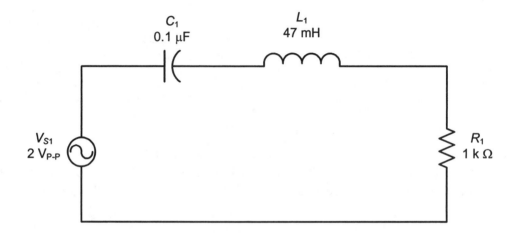

Figure 17-1 Test circuit

3. Change the capacitor to a 0.47µF and repeat the process in step 2, from 500 Hz to 2 kHz in 100 Hz increments.

Questions

1. Plot the resistor voltage magnitude versus frequency for both sets of measurements on the same graph.
2. Plot the resistor voltage phase versus frequency for both sets of measurements on the same graph.
3. What happens to the magnitude and phase shift of the voltage across the resistor when the frequency is below the frequency where the magnitude of X_L equals the magnitude of X_C?

4. What happens to the magnitude and phase shift of the voltage across the resistor when the frequency is above the frequency where the magnitude of X_L equals the magnitude of X_C?
5. What happens to the magnitude and phase shift of the voltage across the resistor at the frequency where the magnitude of X_L equals the magnitude of X_C?
6. Verify KVL for the measurements in step 1.

Title _____ Page No. ___

Signature _____ Witness _____ Date ____

Title _____ Page No. _____

Signature _____ Witness _____ Date _____

131

Title _____ Page No. ____

Signature _____ Witness _____ Date _____

132

Experiment 18: AC Series–Parallel Circuits

Learning Objective

After completing this laboratory experiment, you should be able to:

- Calculate all voltages and currents in single-source AC series–parallel circuits.

Background

The previous experiments examined series and parallel circuits separately with an AC signal source. This experiment combines the series and parallel elements into a single circuit. Techniques that you learned in DC circuits apply to these circuits.

Prelaboratory Preparation

Predict the voltages and currents for each component in the circuits shown in Figures 18-1 and 18-2 for the frequencies specified in the procedure. Generate a results table to record the predictions and the measured results. Also use a circuit simulation software program to verify the predicted results.

Parts List

- 1 kΩ resistor
- 2.2 kΩ resistor
- 0.1 μF capacitor
- 100 mH inductor

Procedure

1. Connect the circuit shown in Figure 18-1. First, set the signal generator to 100 Hz, then 500 Hz, 1 kHz, and 5 kHz. Be sure to adjust the source voltage after each frequency change, if necessary. Measure all the voltage magnitudes and phases. Calculate the current in each branch from the voltages measured.

2. Repeat step 1 for the circuit shown in Figure 18-2.

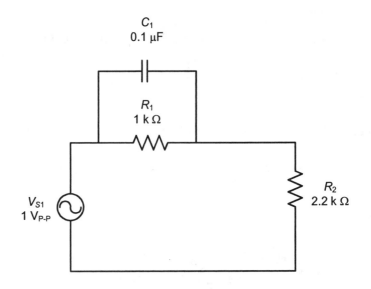

Figure 18-1   Test circuit 1

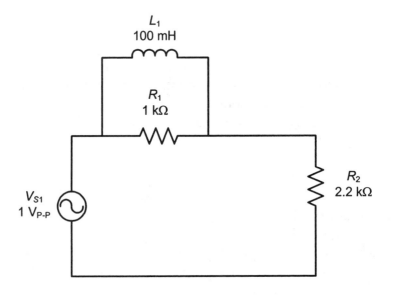

Figure 18-2   Test circuit 2

Questions

1. Explain the change in voltage (magnitude and phase) across R_2 versus frequency for the circuit shown in Figure 18-1.
2. Explain the change in voltage (magnitude and phase) across R_2 versus frequency for the circuit shown in Figure 18-2.
3. What circuit analysis strategy did you verify in this experiment?

Title _____ Page No. __

Signature _____ Witness _____ Date _____

Title _____ Page No. ___

Signature _____ Witness _____ Date _____

Title _____ Page No. ___

Signature _____ Witness _____ Date _____

137

Title _____ Page No. ___

Signature _____ Witness _____ Date _____

138

Experiment 19: Circuit with Both DC and AC Sources

Learning Objective

After completing this laboratory experiment, you should be able to:

- Apply the appropriate circuit analysis techniques for a circuit with both DC and AC sources.

Background

A circuit can have both AC and DC sources.   The AC signal may be superimposed on (added on top of) a DC voltage level. A separate DC analysis and AC analysis must be used to predict the values for this type of circuit.

Prelaboratory Preparation

Predict the voltages across and currents through each passive component, including peak magnitudes, phase shifts, and DC offsets, in Figures 19-1 and 19-2. Generate a results table to record all the results.

Parts List

- 1 kΩ resistor
- (Qty. 2) 2.2 kΩ resistor
- 0.047 µF capacitor
- 100 mH inductor

Procedure

1. **After receiving instructor approval,** connect the circuit shown in Figure 19-1. Use the oscilloscope (DC coupled) to measure the voltage across each resistor in the circuit with all sources activated including any phase shift and DC offset. Calculate the currents through each component from the measured voltages.

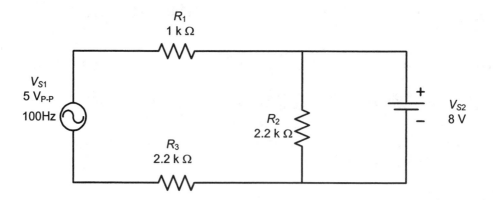

Figure 19-1 Test circuit 1

2. Connect the circuit shown in Figure 19-2. Measure the voltages across each passive component. Use the oscilloscope to make the measurements across the L and C parallel branches. Calculate the currents from the measured voltages.

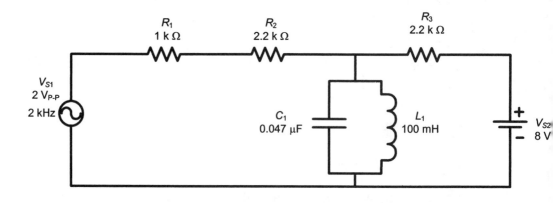

Figure 19-2 Test circuit 2

Questions

1. Compare the peak magnitude, phase shift, and DC offset measurements in both test circuits to predicted values.
2. What was the phase shift of the voltage observed across R_2 in the circuit shown in Figure 19-1? Explain.
3. What was the phase shift of the voltage observed across C_1 in the circuit shown in Figure 19-2? Explain.

Title _____ Page No. __

Signature _____ Witness _____ Date ____

141

Title _____ Page No. ___

Signature _____ Witness _____ Date _____

Title _____ Page No. ___

Signature _____ Witness _____ Date _____

Title _____ Page No. __

Signature _____ Witness _____ Date _____

144

Experiment 20: Complex Power and Power Factor Correction

Learning Objectives

After completing this laboratory experiment, you should be able to:

- Calculate both theoretical and experimental complex power, apparent power, real power, reactive power, power factor angle, and power factor.
- Describe what power factor correction is and why it is important.
- Determine the parallel reactance and component values required for power factor correction.

Background

When power was discussed previously only DC power was examined. What about power in AC circuits? Think about the lights in your home, music from your stereo, and the operation of your computer. These are all cases of AC power. This is also complex power because components other than resistors are in use. In this experiment, the concept of complex power and power factor correction are experimentally verified.

Complex power can always be determined from the general complex power equation:

$$\tilde{S} = \tilde{V}\tilde{I}^* = VI\angle(\theta_V - \theta_I) = S\angle\theta = P + jQ$$

where the voltage and current phasors use RMS values. Complex power for a circuit can also be determined by summing individual component powers (again, with RMS voltage and current phasors). The relevant equations for the R, L, and C components are:

$$R: \; P_R = V_R I_R = \frac{V_R^2}{R} = I_R^2 R \qquad\qquad Q_R = 0$$

$$C: \; P_C = 0 \qquad\qquad Q_C = V_C I_C = \frac{V_C^2}{X_C} = I_C^2 X_C$$

$$L: \; P_L = 0 \qquad\qquad Q_L = V_L I_L = \frac{V_L^2}{X_L} = I_L^2 X_L$$

$$\tilde{S} = \tilde{V}\tilde{I}^* = \sum P + j\left[\sum Q_L - \sum Q_C\right]$$

145

Prelaboratory Preparation

Predict complex power, apparent power, real power, reactive power, power factor angle, and power factor for each circuit shown in the procedure. Determine the parallel reactance and component values required for power factor correction of the circuit shown in Figure 20-1.

Parts List

- 100 Ω resistor
- 47 mH inductor
- 100 mH inductor
- 10 µF capacitor
- 2.2 µF capacitor
- Capacitor for pf correction

Procedure

1. Connect the circuit in Figure 20-1. Measure the voltage magnitude and phase across the inductor (shown with the internal resistance). Calculate the current magnitude and phase. Calculate the experimental complex power, apparent power, real power, reactive power, power factor, and power factor angle.

2. Is the power factor equal to 1? If not, determine the component required to correct the power factor. Place that component in the circuit, in parallel with the inductor, to correct the power factor and repeat step 1.

3. Change the inductor to 100 mH (assume R_{ind} is 100 Ω) and repeat steps 1 and 2.

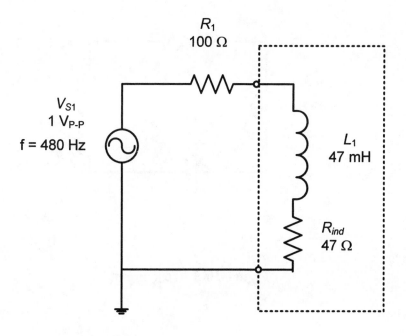

Figure 20-1 Test circuit 1

4. Measure the voltage magnitude and phase across each passive component in the circuits shown in Figure 20-2 and Figure 20-3.

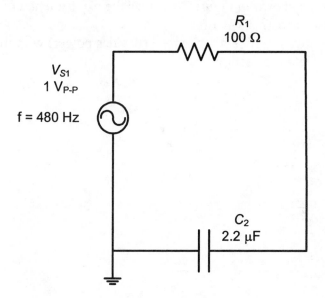

Figure 20-2 Test circuit 2

147

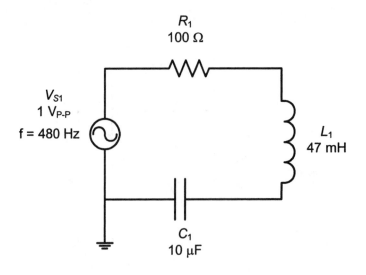

R_1
100 Ω

V_{S1}
1 V$_{P-P}$
f = 480 Hz

L_1
47 mH

C_1
10 μF

Figure 20-3   Test circuit 3

Questions

1. Generate the power triangle before power factor correction for the circuit shown in Figure 20-1 and for the circuit shown in Figure 20-2. Comment on the differences.
2. Generate the power triangle for the circuit shown in Figure 20-3. How would one correct for the power factor of this circuit?
3. In the circuit shown in Figure 20-1, was the experimental power factor correction correct? Explain.
4. What quantity significantly changed (besides power) with the introduction of power factor correction?

Title _____ Page No. __

Signature _____ Witness _____ Date _____

Title _____ Page No. ____

Signature _____ Witness _____ Date _____

Title _____ Page No. _____

Signature _____ Witness _____ Date _____

151

Title _____ Page No. ____

Signature _____ Witness _____ Date _____

Experiment 21: Mesh Analysis

Learning Objective

After completing this laboratory experiment, you should be able to:

- Predict voltages and currents in DC and AC circuits using the mesh circuit analysis technique.

Background

As circuits become more complex, series—parallel analysis techniques such as superposition become difficult to apply. More advanced techniques are needed. A technique to analyze complex circuits is mesh analysis. This experiment has three circuits upon which you will perform a mesh analysis and then confirm all calculated values with measured values. A comparison of the theoretical, experimental, and simulation results will be made.

Prelaboratory Preparation

Predict the voltages across and currents through each component using the mesh analysis circuit technique for each circuit shown in the procedure. Also use a circuit simulation program to verify the predicted results.

Parts List

- 33 Ω resistor
- 47 Ω resistor
- 220 Ω resistor
- 330 Ω resistor
- 680 Ω resistor
- (Qty. 2) 2.2 kΩ resistors
- 6.8 kΩ resistor
- 8.2 kΩ resistor
- 47 mH inductor
- 100 mH inductor

Procedure

1. In the circuit shown in Figure 21-1, measure the voltages across and the currents through each of the resistors. Calculate the power dissipated by each resistor.

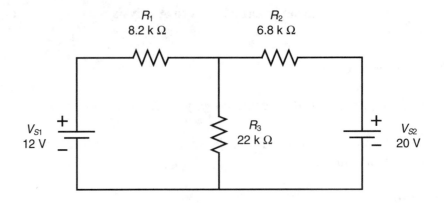

Figure 21-1 Test circuit 1

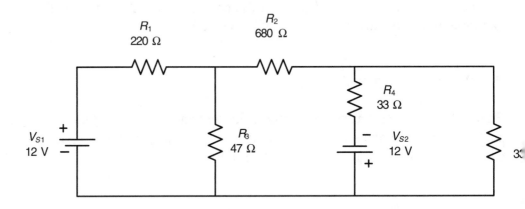

Figure 21-2 Test circuit 2

2. Repeat the process for the circuit shown in Figure 21-2.

3. For the circuit shown in Figure 21-3, with a frequency of 5 kHz, measure the voltage magnitude and phase across each component. *Hint*: Use circuit rearrangement.

154

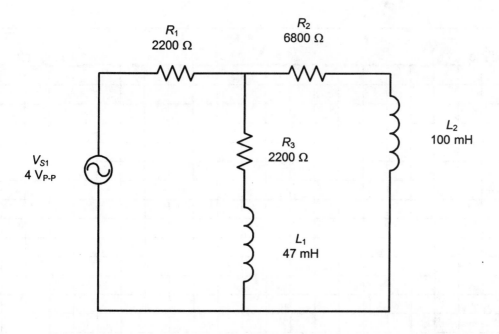

Figure 21-3   Test circuit 3

Questions

1. What was the percentage difference between predicted voltages and currents to measured voltages and currents for the components in the circuit shown in Figure 21-1? In Figure 21-2? In Figure 21-3?
2. What law(s) was verified for DC circuit analysis in this experiment?
3. What law(s) was verified for AC circuit analysis in this experiment?
4. Determine the mesh currents for the circuit in Figure 21-3 from the appropriate voltage measurements. How do they compare to the mesh current predictions?
5. What is the effect due to the inductors on the voltage measurements in Figure 21-3? Explain.
6. How would the analysis of the circuit shown in Figure 21-2 change if R_4 were shorted? Show that analysis.
7. Compare the results for the circuit shown in Figure 21-1 with the circuit shown in Figure 11-3. Explain any significant differences.

Title _____ | Page No. ____

Signature _____ Witness _____ Date _____

Title _____ Page No. __

Signature _____ Witness _____ Date _____

157

Title _____ Page No. ____

Signature _____ Witness _____ Date _____

Title _____ Page No. ____

Signature _____ Witness _____ Date _____

159

Title _____ Page No. _____

Signature _____ Witness _____ Date _____

Experiment 22: Nodal Analysis

Learning Objective

After completing this laboratory experiment, you should be able to:

- Predict voltages and currents in DC and AC circuits using the nodal (node voltage) analysis technique.

Background

Another technique for analysis of complex circuits is nodal analysis. Three circuits are examined in this experiment using nodal analysis. The quantities will be verified with measurements. A comparison to the results of Experiment 21 is expected.

Recall that mesh analysis utilized KVL to solve for mesh currents. Nodal analysis utilizes KCL to solve for node voltages.

Prelaboratory Preparation

Use the nodal analysis technique to determine the voltages across and currents through each component in the circuits shown in Figures 22-1, 22-2, and 22-3. Also use a circuit simulation program to verify your predicted results. Compare your theoretical results to those of Experiment 21.

Parts List

- 33 Ω resistor
- 47 Ω resistor
- 220 Ω resistor
- 330 Ω resistor
- 680 Ω resistor
- 8.2 kΩ resistor
- (Qty. 2) 2.2 kΩ resistors
- 6.8 kΩ resistor
- 47 mH inductor
- 100 mH inductor

Procedure

Construct each circuit and measure the component voltages and node voltages in all circuits. Measure the component currents in the DC circuits. Determine experimental currents through components from the voltage measurement in the AC circuit. The frequency for the AC source is 5 kHz.

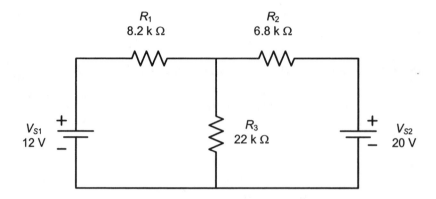

Figure 22-1 Test circuit 1

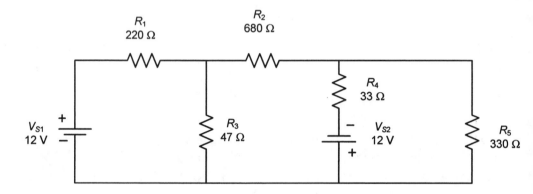

Figure 22-2 Test circuit 2

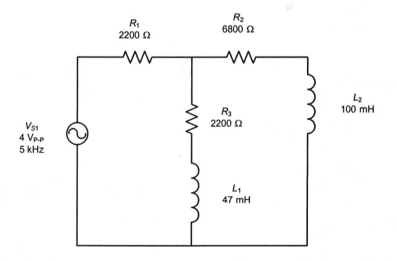

Figure 22-3 Test circuit 3

162

Questions

1. Show percentage differences from theoretical in the results table for each circuit. Explain any significant differences.
2. What is the effect of the inductors in Figure 22-3? Explain.
3. How do the experimental results for the three circuits compare with those in Experiment 21?
4. What law(s) is verified for DC circuit analysis in this experiment?
5. What law(s) is verified for AC circuit analysis in this experiment?
6. How would the analysis of the circuit shown in Figure 22-2 change if R_4 were shorted? Show that analysis.
7. Compare the results for the circuit shown in Figure 21-1 with the circuit shown in Figure 11-3. Explain any significant differences.

Title _____ Page No. _____

Signature _____ Witness _____ Date _____

Title _____ Page No. ___

Signature _____ Witness _____ Date _____

165

Title _____ Page No. _____

Signature _____ Witness _____ Date _____

Title _____ Page No. ___

Signature _____ Witness _____ Date ___

167

Title

Page No.

Signature   Witness   Date

Experiment 23: Bridge Circuits and Delta–Wye Conversions

Learning Objectives

After completing this laboratory experiment, you should be able to:

- Explain why bridge circuits are used and determine if a bridge circuit is balanced.
- Perform delta–wye conversions.

Background

The bridge circuit can be used to make very precise measurements of resistance, temperature, and so on. The balanced bridge can be described by a proportionality of two resistor pairs. When the ratio of the two resistor pairs is equal, then no current flows in the middle of the bridge. See Figure 23-1 as an example of a bridge circuit.

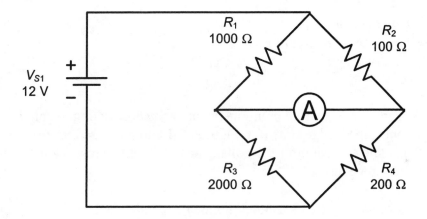

Figure 23-1 Balanced bridge example

If the ratio of $R_1/R_3 = R_2/R_4$ then the current through the current meter will be 0 A. To use this as a measuring circuit, replace R_1 with a variable resistance and R_3 with the resistance under test. When R_1 is adjusted to null the current, the value of R_3 can be found.

The delta–wye circuit conversions are useful because they simplify the analysis in some circuits. It may be easier to analyze a circuit with one configuration versus the other. Delta–wye circuits are used extensively in three-phase systems. A wye circuit can be shown as a T and the delta circuit can be shown as a pi.

Prelaboratory Preparation

In your engineering notebook explain the operation of a balanced bridge. Use a circuit simulation software program to verify the operation of the balanced bridge shown in Figure 23-1.

Predict the voltage across terminals 3 and 4 for the circuit shown in Figure 23-2. Convert the circuit shown in Figure 23-2 to a delta configuration. Then predict the voltage across terminals 3 and 4 assuming that a 10 VDC source is placed across terminals 1 and 2 with the positive source terminal connected to terminal 1. Also use a circuit simulation program to verify the predictions.

Parts List

- $100 \, \Omega$ resistor
- $200 \, \Omega$ resistor
- (Qty. 2) $1 \, k\Omega$ resistors
- $2 \, k\Omega$ resistor
- $5 \, k\Omega$ potentiometer
- $10 \, k\Omega$ resistor
- (Qty. 2) $50 \, k\Omega$ potentiomenter

Procedure

1. Be sure to measure all components prior to constructing the circuit. Connect the bridge shown in Figure 23-1 and measure the current through the ammeter. Monitor the voltage across R_3 and across R_4 during the entire procedure.

2. Replace R_1 with a $5 \, k\Omega$ potentiometer and balance the bridge. Replace R_3 with a $50 \, k\Omega$ potentiometer and set R_3 to 3 different values, balancing the bridge each time, with R_1. With the bridge balanced, measure the resistance of R_1 and predict the value of R_3. Now measure R_3 and record these measures in the results section of your engineering notebook.

3. With the bridge still balanced, and the power turned off, replace the ammeter with a $1 \, k\Omega$ resistor. Turn the power back on and monitor the voltage across R_3 and across R_4 for any changes from the previous step.

4. With the bridge still balanced, replace the $1 \, k\Omega$ resistor with an open and monitor the voltage across R_3 and across R_4 for any changes from the previous step. Now replace with a short and monitor the voltage across R_3 and across R_4 for any changes from the previous step. See Question 2.

5. Connect the circuit shown in Figure 23-2. Measure the resistance across terminals 3 and 4 with the power supply removed and left as an open.

Measure the resistance across terminals 3 and 4 with the power supply removed and replaced as a short circuit. Remove the short, reinstall the supply, apply power, and measure the voltage across terminals 3 and 4.

6. Connect the equivalent delta circuit that you predicted. Repeat the measurements made on the wye circuit.

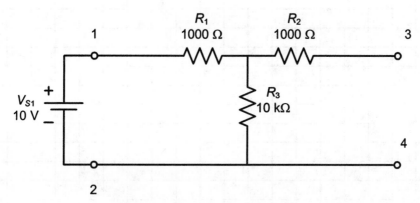

Figure 23-2 Test circuit 2

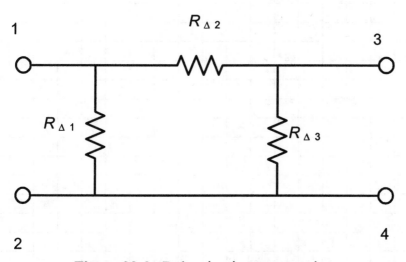

Figure 23-3 Delta circuit representation

Questions

1. When the bridge was first connected was the current zero? Explain why or why not.
2. Was there any voltage change across R_3 and R_4 when the ammeter was changed to 1 kΩ or to an open or to a short? Explain.
3. Compare the predicted value of V_{R3} with the measured value of V_{R3} in Step 1. Explain any differences.
4. Are the measurements for the delta and wye circuits the same? Should these be the same? Explain.

Title

Page No.

Signature _____ Witness _____ Date _____

Title _____ Page No. ___

Signature _____ Witness _____ Date _____

173

Title		Page No.

Signature	Witness	Date

Experiment 24: AC Steady–State Response of an RLC Circuit

Learning Objectives

After completing this laboratory experiment, you should be able to:

- Predict voltages and currents in AC circuits using the nodal (node voltage) analysis circuit technique.
- Predict voltages and currents in AC circuits using the mesh analysis circuit technique.
- Predict voltages and currents in AC circuits using delta–wye conversions.

Background

This experiment examines the operation of an RLC circuit at two frequencies. The analysis techniques learned in earlier experiments are applied here.

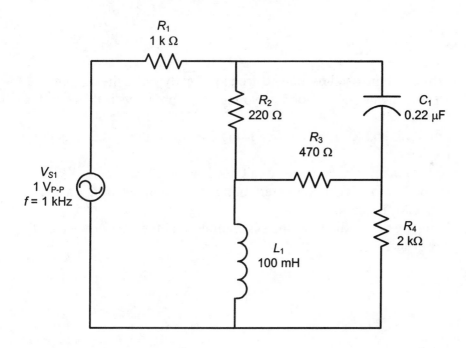

Figure 24-1   RLC test circuit

Prelaboratory Preparation

Prior to the laboratory session predict the voltages and currents at 1 kHz for each of the components in the circuit shown in Figure 24-1, using both nodal and mesh analysis and delta–wye conversion with series–parallel circuit analysis. Verify

175

the results using a circuit simulation program. Run the circuit simulation again at 10 kHz.

Parts List

- 220 Ω resistor
- 470 Ω resistor
- 1 kΩ resistor
- 2 kΩ resistor
- 0.22 µF capacitor
- 100 mH inductor

Procedure

Be sure to measure all components prior to constructing the circuit. Construct the circuit in Figure 24-1. Measure and record all node voltages including phases at 1 kHz. Measure and record all component voltage and current magnitudes using a DMM. See Question 1. Set the signal generator to 10 kHz and repeat all measurements.

Questions

1. Why can't the oscilloscope be used to directly make the current measurements and component voltage measurements if the circuit is not rearranged?
2. Can KCL and KVL be verified using the DMM magnitude results? Explain.
3. At how many points in the circuit does the oscilloscope need to be placed to obtain enough data for determining all component voltages and currents? Explain.
4. Compare the predictions, the simulations, and the measurements. Explain any differences.

Title _____ Page No. ___

Signature _____ Witness _____ Date _____

Title

Page No.

Signature _____ Witness _____ Date _____

Title _____ Page No. _____

Signature _____ Witness _____ Date _____

Title _____ Page No. ___

Signature _____ Witness _____ Date _____

Experiment 25: Thevenin Equivalent Circuits

Learning Objectives

After completing this laboratory experiment, you should be able to:

- Explain what an equivalent circuit is and why equivalent circuits are needed.
- Determine the Thevenin equivalent circuit of a given test circuit.
- Predict the voltage and current for a given load using the Thevenin equivalent circuit.
- Determine the load resistance required for maximum power transfer.

Background

The Thevenin equivalent circuit is a model that can be used to predict the performance of a circuit when a load may be changed. The Thevenin equivalent circuit will not change as the load is changed. This allows one to predict the output using the voltage divider. The Thevenin equivalent circuit is applicable at only one frequency in circuits with reactances.

Prelaboratory Preparation

Predict the Thevenin equivalent circuit for the circuits shown in Figures 25-1 and 25-2. For both circuits predict the load voltage and current using load resistors of 1 kΩ, 6.8 kΩ, and 10 kΩ. For both circuits predict the required load resistance to obtain maximum power transfer and predict the associated load voltage and load current. Repeat the analysis for the AC circuit at 3 kHz.

Parts List

- (Qty. 2) 1 kΩ resistor
- 3.3 kΩ resistor
- 4.7 kΩ resistor
- 6.8 kΩ resistor
- 10 kΩ resistor
- 5 kΩ potentiometer
- 0.1 μF capacitor
- 100 mH inductor
- Resistors for maximum power transfer

Procedure

1. Be sure to measure all components prior to constructing the circuit. Construct the circuit shown in Figure 25-1. Measure the Thevenin voltage and the Thevenin resistance. Connect each load and measure the load voltage and load current. For comparison, construct the Thevenin equivalent circuit that was predicated in the prelaboaratory preparation. Connect each load and measure the associated load voltage and current. Report any significant differences.

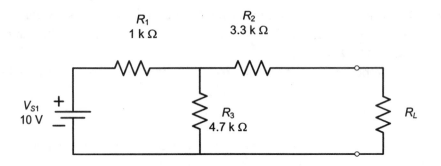

Figure 25-1 DC test circuit

2. Then connect the load that will provide maximum power transfer and verify the correct operation.

3. Construct the circuit shown in Figure 25-2. Again measure all components to ensure correct values. Measure the output voltage, amplitude and phase, with no load attached, at 300 Hz and 3 kHz.

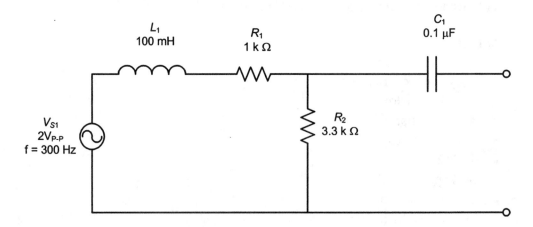

Figure 25-2 AC test circuit

4. Connect a short circuit to the output and by some appropriate means determine the magnitude and phase of the short circuit current for both frequencies. *Hint*: Could the voltage across the capacitor help? Use these values to determine the Thevenin equivalent impedance at each frequency.

$$\tilde{Z}_{TH} = \frac{\tilde{V}_{OC}}{\tilde{I}_{SC}}$$

5. Connect each load to the output terminals and measure the output voltage amplitude and phase for both frequencies.

6. Repeat for the load resistor that will achieve maximum power transfer. Do this for both frequencies.

7. Be sure to measure the DC resistance of the inductor.

Questions

1. Compare the theoretical results to the measured results. Explain any significant differences.
2. At which of the two frequencies of the AC circuit would the system be capable of transmitting more power to the load and why? Do the analysis for both the 1 kΩ load and the resistive load for maximum power transfer load.
3. What circuit analysis concept is verified in this experiment? Why is it useful?

4. Why is $\tilde{Z}_{TH} = \tilde{V}_{OC}/\tilde{I}_{SC}$ valid? *Hint*: Consider a source conversion.
5. Did the equivalent circuits and load results for the AC circuit differ between 300 Hz and 3 kHz? Explain.

Title _____ Page No. ___

Signature_____ Witness_____ Date_____

Title _____ Page No. ____

Signature _____ Witness _____ Date _____

185

Title _____ | Page No. ____

Signature _____ Witness _____ Date ____

Title _____ Page No. __

Signature _____ Witness _____ Date _____

187

Title _____ Page No. _____

Signature _____ Witness _____ Date _____

Experiment 26: Norton Equivalent Circuits

Learning Objectives

After completing this laboratory experiment, you should be able to:

- Explain what an equivalent circuit is and why equivalent circuits are needed.
- Determine the Norton equivalent circuit of a given circuit.
- Predict the voltage and current for a given load using the Norton equivalent circuit.
- Determine the load resistance required for maximum power transfer.

Background

The Norton equivalent circuit is a model that can be used to predict the performance of a circuit when a load may be changed. The Norton equivalent circuit will not change as the load is changed. This allows one to predict the output using the current divider. The Norton equivalent circuit, like the Thevenin equivalent circuit, is applicable at only one frequency in circuits with reactances.

Prelaboratory Preparation

Predict the Norton equivalent circuit for the circuits shown in Figures 26-1 and 26-2. For both test circuits predict the voltage and current at the load using load resistors of 1 kΩ, 6.8 kΩ, and 10 kΩ using the Norton equivalent circuit. Record the results in a data table in the Results section of the lab report. For both test circuits predict the required load resistance to obtain maximum power transfer and predict the corresponding load voltage and load current. Repeat the analysis for the circuit shown in Figure 26-2 at 3 kHz.

Parts List

- (Qty. 2) 1 kΩ resistor
- 3.3 kΩ resistor
- 4.7 kΩ resistor
- 6.8 kΩ resistor
- 10 kΩ resistor
- 5 kΩ potentiometer
- 0.1 μF capacitor
- 100 mH inductor
- Resistors for maximum power transfer

189

Procedure

1. Construct the circuit shown in Figure 26-1 and measure the quantities needed to determine the Norton equivalent circuit. Connect each load and measure the load voltage and current.

2. Connect the load resistor that will provide maximum power transfer and measure the load voltage and current. Calculate the power at the load.

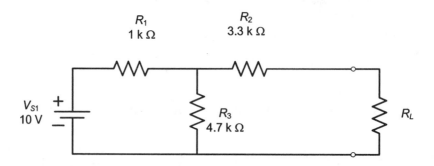

Figure 26-1 Test circuit 1

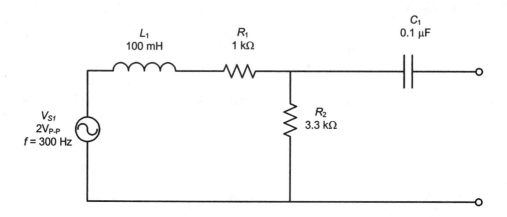

Figure 26-2 Test circuit 2

3. Repeat steps 1 and 2 for Figure 26-2 at 300 Hz and 3 kHz.

Questions

1. Compare the theoretical results to the measured results. Explain any significant differences.
2. Did frequency affect the Norton equivalent circuit? Why?

3. What circuit analysis concept is verified in this experiment? Why is it useful?

4. Why is $\tilde{Z}_{TH} = \tilde{V}_{OC} / \tilde{I}_{SC}$ valid? *Hint*: Consider a source conversion.

5. Did the equivalent circuits and load results for the AC circuit differ between 300 Hz and 3 kHz? Explain.

6. Compare the results of this experiment with the results of Experiment 25. Source conversions are useful here to compare the equivalent circuits. Compare load voltage and current results in a table.

Title _____ Page No. ____

Signature _____ Witness _____ Date _____

Title _____ Page No. _____

Signature _____ Witness _____ Date _____

Title _____ Page No. ___

Signature _____ Witness _____ Date _____

194

Title _____ Page No. __

Signature _____ Witness _____ Date _____

195

Title _____ Page No. ____

Signature _____ Witness _____ Date

196

Experiment 27: Maximum Power Transfer

Learning Objectives

After completing this laboratory experiment, you should be able to:

- Determine the load resistance required for maximum power transfer.
- Determine the trend in power transfer from the source to the load as the load is varied.

Background

To fully utilize the energy of the circuit, one must transfer the maximum power from the source to the load. You will demonstrate the maximum power transfer theorem in this experiment.

The maximum power transfer theorem states that the maximum power is transferred from the source to the load when the source resistance matches the load resistance in a resistive circuit.

Prelaboratory Preparation

Predict the voltage, current, and power at the load for the circuits shown in Figures 27-1 and 27-2 using the loads indicated in the procedure. Construct a results table to record all data required for the procedure. Plot the theoretical results of power versus load resistance. Predict the load for maximum power transfer and the power for that load using an equivalent circuit.

Parts List

- 100 Ω resistor
- (Qty. 2) 220 Ω resistors
- 330 Ω resistor
- (Qty. 2) 470 Ω resistors
- 560 Ω resistor
- 680 Ω resistor
- 820 Ω resistor
- (Qty. 2) 1 kΩ resistors
- 1.2 kΩ resistor
- 1.5 kΩ resistor
- 2 kΩ resistor
- 2 kΩ potentiometer

96.876 Ω
218.315Ω
217.472 Ω
327.390 Ω Large
462.207 Ω
477.264 Ω
565.604 Ω
679.748
834.241Ω 834.261Ω
988.693Ω
983.485Ω
1.1846 kΩ
1.4434 k
1.9684 kΩ

197

Procedure

1. Connect the circuit in Figure 27-1. Use a 2 kΩ potentiometer for R_L. Start with the potentiometer at the lowest setting and measure the current through and the voltage across the potentiometer. Increase the potentiometer in approximately 200 Ω increments and continue to monitor the current and voltage. Continue this to the maximum setting of the potentiometer. Also set the potentiometer to the predicted value for maximum power transfer. Make the necessary measurements.

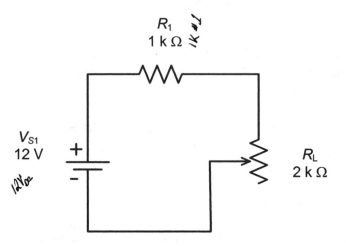

Figure 27-1   Test circuit 1

2. Now replace the potentiometer with the following fixed resistors: 100 Ω, 220 Ω, 330 Ω, 470 Ω, 560 Ω, 680 Ω, 820 Ω, 1 kΩ, 1.2 kΩ, 1.5 kΩ, and 2 kΩ. Monitor current and voltage for each case.

3. Connect Figure 27-2. Use a 2 kΩ potentiometer for R_L. Start with the potentiometer at the lowest setting and measure the current through and the voltage across the potentiometer. Increase the potentiometer in 200 Ω increments and continue to monitor the current and voltage. Continue this to the maximum setting of the potentiometer. Also set the potentiometer to the predicted value for maximum power transfer. Make the necessary measurements.

198

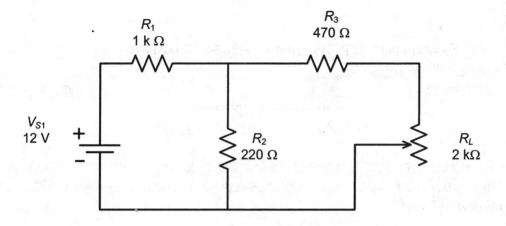

Figure 27-2 Test circuit 2

4. Now replace the potentiometer with the following fixed resistors, 100 Ω, 220 Ω, 330 Ω, 470 Ω, 560 Ω, 680 Ω, 820 Ω, 1 kΩ, 1.2 kΩ, 1.5 kΩ , and 2 kΩ. Monitor current and voltage for each case.

5. Replace the DC source with a 1 $V_{p\text{-}p}$ AC and repeat Step 4.

Questions

1. Calculate the power dissipated for each potentiometer setting in step 1. Plot the power on the *y*-axis and potentiometer setting on the *x*-axis. On the same graph, plot predicted power and measured power versus load resistance of the fixed resistors for step 2. Explain any significant differences.
2. Calculate the power dissipated for each potentiometer setting in step 3. Plot the power on the *y*-axis and potentiometer setting on the *x*-axis. On the same graph, plot predicted power and measured power versus load resistance of the fixed resistors for step 4. Explain any significant differences.
3. Is there a difference in circuit behavior between applying the AC source and the DC source? Explain.
4. Would maximum power transfer hold true for loads with reactance? Explain.
5. What is the trend in power transfer versus load resistance?
6. Did the measurements confirm the prediction on the loads for maximum power transfer? Explain.

ROBERT J. JOHNSON Jr

ELCTEC-124   02/11/2010

BACKGROUND INFORMATION

TO FULLY UTILIZE THE ENERGY OF THE CIRCUIT, ONE MUST TRANSFER THE MAXIMUM POWER FROM THE SOURCE TO THE LOAD. ~~YOU WILL BE~~ ~~DEMONSTRATE~~ BY

PROCEDURE

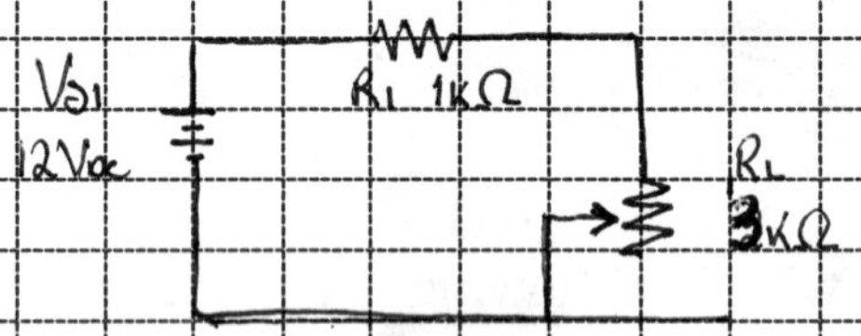

FIGURE 27-1

WIRE THE CIRCUIT AS SEEN IN FIGURE 27-1. USE A 3KΩ POTENTIOMETER FOR THE LOAD. WITH THE POTENTIOMETER SET TO 0Ω, MEASURE THE VOLTAGE ACROSS THE POTENTIOMETER AS WELL AS THE CURRENT THROUGH IT. INCREASE THE POTENTIOMETER IN 200Ω INCREMENTS AND RECORD THE VOLTAGE AND CURRENT IN A TABLE. BE SURE TO RECORD THE MAXIMUM POTENTIOMETER VALUES AS WELL AS THE PREDICTED MAXIMUM POWER TRANSFER VALUES.

MEASURE AND RECORD THE RESISTANCE VALUES FOR THE FOLLOWING RESISTORS: 100Ω, 200Ω, 330Ω, 470Ω, 560Ω, 680Ω, 820Ω, 1KΩ, 1.2KΩ, 1.5KΩ AND 2KΩ. REPLACE THE POTENTIOMETER WITH EACH OF THE RESISTORS LISTED. MEASURE AND RECORD THE VOLTAGE AND CURRENT FOR EACH RESISTOR.

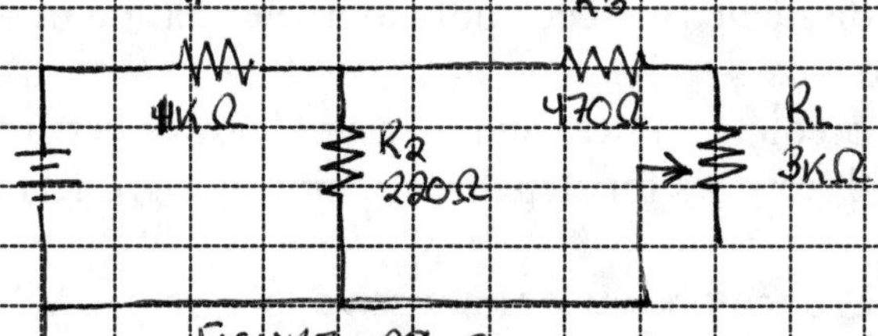

FIGURE 27-2

WIRE THE CIRCUIT AS SEEN IN FIGURE 27-2. USE A 3KΩ POTENTIOMETER FOR THE LOAD. AGAIN, MEASURE AND RECORD THE VOLTAGE AND CURRENT VALUES FOR THE POTENTIOMETER FOR ALL 50 VALUE RESISTANCES FROM 0Ω TO MAXIMUM OHMS. DO THIS IN 200Ω INCREMENTS. ALSO RECORD THE PREDICTED VALUES FOR MAXIMUM POWER TRANSFER.

Signature _____ Witness _____ Date 2-18-10

200

ROBERT J. JOHNSON JR

ELCTEC 124   02/11/2010

PROCEDURE - CONTINUED

USING THE RESISTORS FROM BEFORE, REPLACE THE POTENTIOMETER WITH
EACH OF THE RESISTOR VALUES AND RECORD THE VOLTAGE AND CURRENT
FOR EACH RESISTOR VALUE

REPLACE THE DC VS1 WITH A 1VPP AC SOURCE AND REPEAT THE
PREVIOUS STEPS.   DI-B5
                                 DATA

MEASURED RESULTS

CIRCUIT ONE

RESISTANCE	VOLTAGE	CURRENT (mA)	RESISTANCE	VOLTAGE	CURRENT (mA)
0 Ω	.74 V	12.121	(100) 98.876 Ω	1.110 V	12.130
200 Ω	2.022 V	10.068	(220) 218.315 Ω	2.165 V	9.922
400 Ω	3.408 V	8.663	(330) 327.390 Ω	2.977 V	9.096
600 Ω	4.504 V	7.556	(470) 462.207 Ω	3.813 V	8.250
800 Ω	5.349 V	6.697	(560) 565.604 Ω	4.355 V	7.698
1000 Ω	6.019 V	6.022	(670) 672.768 Ω	4.843 V	7.206
1200 Ω	6.473 V	5.563	(820) 834.214 Ω	5.477 V	6.569
1400 Ω	7.023 V	5.008	(1K) 983.485 Ω	5.965 V	6.073
1600 Ω	7.378 V	4.650	(1.2K) 1.1846 kΩ	6.523 V	5.511
1800 Ω	7.711 V	4.316	(1.5K) 1.4934 kΩ	7.202 V	4.827
2000 Ω	8.012 V	4.012	(2K) 1.9684 kΩ	7.969 V	4.054

CIRCUIT TWO

RESIST	VOLTS	CURRENT	RESISTANCE	VOLTS (DC)	IDC (mA)	VOLTS (AC)	IAC
0 Ω	14.719 mV	3.285 mA	98.876 Ω	.2946	2.865	16.017 mV	112 μA
200 Ω	492.3 mV	2.564 A	218.315 Ω	.5533	2.469	30.748 mV	12 μA
400 Ω	826 mV	2.055 A	327.390 Ω	.7325	2.196	40.902 mV	11 μA
600 Ω	1.037 mV	1.733 A	462.207	.9061	1.933	50.715 mV	10 μA
800 Ω	1.193 V	1.495 A	565.604	1.012	1.770	56.762 mV	10 μA
1K Ω	1.316 V	1.305 A	672.768	1.106	1.628	62.052 mV	11 μA
1.2K Ω	1.407 V	1.168 A	834.214	1.221	1.452	68.581 mV	9 nA
1.4K Ω	1.486 V	1.047 A	983.485	1.306	1.320	134.81 mV	8 μA
1.6K Ω	1.535 V	972.3 mA	1.1846 K	1.401	1.176	124.73 mV	7 μA
1.8K Ω	1.588 V	894.4 mA	1.4934 K	1.511	1.007	111.98 mV	8 μA
2.0K Ω	1.635 V	818.1 mA	1.9684 K	1.604	.8259	46.722 mV	4 nA

Signature _____   Witness _____   Date 2-11-10

Title	EXPERIMENT 27	MAXIMUM POWER TRANSFER			Page No. 3

ROBERT J. JOHNSON Jr

ELCTEC-124   02/11/2010

THEORETICAL RESULTS ~~THEORETICAL RESULTS BY~~

CIRCUIT ONE

RESISTANCE	VOLTAGE	CURRENT (mA)	RESISTANCE	VOLTAGE	CURRENT (mA)
0 Ω	0.000V	12.0 mA	100 Ω	1.091 V	10.909 mA
200 Ω	2.000V	10.0 mA	220 Ω	2.164 V	9.836 mA
400 Ω	3.429V	8.571 mA	330 Ω	2.977 V	9.023 mA
600 Ω	4.500V	7.500 mA	470 Ω	3.837 V	8.163 mA
800 Ω	5.333V	6.667 mA	560 Ω	4.308 V	7.692 mA
1K Ω	6.000V	B2 5.909 mA / 5.433 mA	670 Ω	4.814 V	7.180 mA
1.2K Ω	6.545V	5.455 mA	820 Ω	5.407 V	6.593 mA
1.4K Ω	7.000V	B5 4.615 mA	1K Ω	6.000 V	6.000 mA
1.6K Ω	7.385V	4.616 mA	1.2K Ω	6.245V	5.455 mA
1.8K Ω	7.714V	4.286 mA	1.5K Ω	7.200 V	4.800 mA
2.K Ω	8.000V	4.000 mA	2.0K Ω	8.000 V	4.000 mA

CIRCUIT TWO

RESIST	VOLTS	CURRENT	RESIST	VOLTS(Ω)	I_OC (mA)	VOLTS (AC) V	I_OC
0 Ω	0.000V	3.327 mA	100 Ω	288.4 mV	2.884	16.991	170 μA
200 Ω	510 mV	2.545 mA	220 Ω	547.0 mV	2.486	32.027	146 μA
400 Ω	824 mV	2.060 mA	330 Ω	728.4 mV	2.207	42.917	130 μA
600 Ω	1.038 V	1.731 mA	470 Ω	907.8 mV	1.932	53.485	114 μA
800 Ω	1.194 V	1.442 mA	560 Ω	1.001 V	1.788	58.988	108 μA
1000 Ω	1.311 V	1.311 mA	670 Ω	1.098 V	1.639	64.696	97 μA
1200 Ω	1.403 V	1.164 mA	820 Ω	1.207 V	1.472	71.102	87 μA
1400 Ω	1.478 V	1.053 mA	1K Ω	1.311 V	1.311	77.259	77 μA
1600 Ω	1.539 V	.962 mA	1.2K Ω	1.403 V	1.169	82.683	64 μA
1800 Ω	1.590 V	.883 mA	1.5K Ω	1.509 V	1.006	88.934	54 μA
2000 Ω	1.633 V	.816 mA	2.0K Ω	1.633 V	.816	96.208	48 μA

SAMPLE CALCULATIONS

CIRCUIT ONE   CIRCUIT TWO

$R_{TH} = R_1 || R_2 + R_3$

MAX POWER = $R_L = R_1$ $R_{TH} = 650.328 \, \Omega$

$R_L = 1000 \, \Omega$ MAX POWER = $R_L = R_{TH}$

$P = I^2 R \Rightarrow P = (6 mA)^2 (1K\Omega) = 36 mW$ $P = (1.639 mA)(670 \Omega) = 18 mW$

Signature		Witness		Date 2-11-10

Title Experiment 27 maximum power transfer   Page No. 4

Robert J. Johnson Jr

ELCTEC-124   02/11/2010

ANALYSIS OF RESULTS

Signature _____ Witness _____ Date _____

Title _____ Page No. ___

Signature _____ Witness _____ Date _____

204

Experiment 28: Transformers

Learning Objective

After completing this laboratory experiment, you should be able to:

- Predict voltages, currents, and impedances in circuits that contain a transformer.

Background

The transformer can be used to magnetically couple electrical energy from one portion of the circuit to another portion of the circuit. The transformer can be a step up, step down, or isolation transformer. You will examine the operation of the transformer and the equations used for analysis. Then you will construct circuits using transformers and make the proper measurements.

Prelaboratory Preparation

Prior to the laboratory session predict the primary current, the secondary voltage, and the input impedance for the circuit shown in Figure 28-1 for each of the following loads at a frequency of 2 kHz:

1. No load (open circuit)
2. 100 Ω resistor
3. 0.47 μF capacitor
4. 47 mH inductor

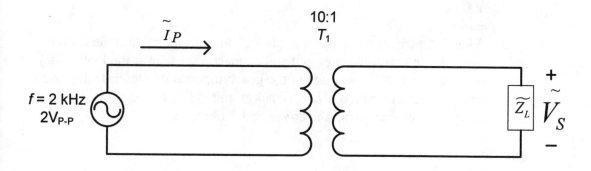

Figure 28-1 Transformer test circuit

Parts List

- Transformer 10:1
- (Qty. 2) 100 Ω resistors
- 0.47 µF capacitor
- 47 mH inductor

Procedure

1. Construct the circuit in Figure 28-2. Measure all component values for acceptable readings. Connect each of the four loads. Record the secondary voltage and primary current. These should be magnitude and phase measurements with the signal generator voltage as a reference.

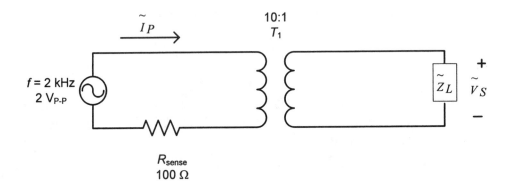

Figure 28-2   Test circuit with 100 Ω sense resistor

2. For the 100 Ω load, also measure the secondary voltage at 15 Hz and 100 kHz.

3. After completing all measurements with all four loads, turn the signal generator to a minimum output and connect a speaker as the load. Vary the frequency and amplitude of the generator output. Describe any sound changes versus frequency and versus amplitude. Change from sine wave to square wave and describe any sound difference.

Questions

1. Determine the experimental input impedance to the transformer using the measurements for each load. Compare all experimental results to the theoretical results and explain any differences.
2. Explain the differences in the secondary voltages between the different loads.

3. Did the transformer operate at 15 Hz and 100 kHz as it did at 2 kHz? Explain. Does this imply that the transformer is frequency dependent?
4. Explain the reasons behind the observations when the speaker was connected.

Title _____ Page No. ___

Signature _____ Witness _____ Date

Title _____ Page No. ___

Signature _____ Witness _____ Date _____

Title _____ Page No. _____

Signature _____ Witness _____ Date _____

210

Experiment 29: Three-Phase Circuits

Learning Objective

After completing this laboratory experiment, you should be able to:

- Predict current and voltage magnitudes in a three-phase circuit.

Background

Three-phase is the type of power transmission system that has three signals with a special phase relationship with respect to each other. There are three AC sinusoidal steady-state signals with the same frequency and peak magnitude but different phases, usually 120° apart. The "phase" in three-phase indicates that the signals have different phases. The instantaneous power is constant in a three-phase system. This becomes a tremendous advantage with large motors and industrial machinery. Why? The transmission of electrical energy over a three-phase system requires only three lines as compared to three single-phase systems, which require six lines.

Prelaboratory Preparation

Predict the voltage and current magnitudes for a three-phase load with 1 kΩ resistors shown in Figure 29-1. Ask your instructor what three-phase source voltage is to be used in this experiment.

Parts List

- (Qty. 3) 1 kΩ resistors instructor will indicate the power rating
- (Qty. 3) Single-phase, insulated winding 10:1 transformers
- (Qty. 3) 1N4001 diodes

Procedure

(Prior to performing this experiment, check with your instructor for the proper setup and safety procedures in your laboratory)

1. In this laboratory experiment single-phase transformers are used. Be sure to read the faceplate data/specifications of the transformers used in the laboratory. Do *not* exceed the ratings of the transformers at any time during the procedure.

2. Connect the transformers so that a wye configuration is obtained for the source.

3. Connect the load in a wye configuration as shown in Figure 29-1 and measure the secondary current and voltages for each phase. Are the three phase voltages 120° apart?

4. Replace the three 1 kΩ resistors with diodes and insert a 1 kΩ load, as shown in Figure 29-2. Measure the DC voltage across R_1. Using an oscilloscope, sketch the voltage across R_1. The DC voltage across R_1 should measure:

$$V_{DC} = 0.318\left(V_{peak\ phase} - 0.7\right)$$

5. Disconnect two diodes and sketch the output waveform across R_1. It should resemble a half-wave rectifier.

6. Remove power and add a second diode to the circuit. Apply power and sketch the voltage across R_1. Record the DC voltage across R_1.

7. Remove power and add the third diode to the circuit. Measure the frequency of the ripple voltage across R_1.

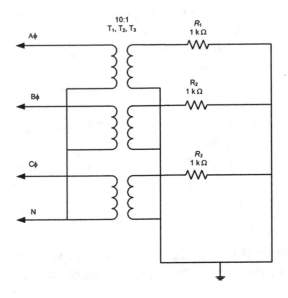

Figure 29-1 Resistor load configuration for three-phase experiment

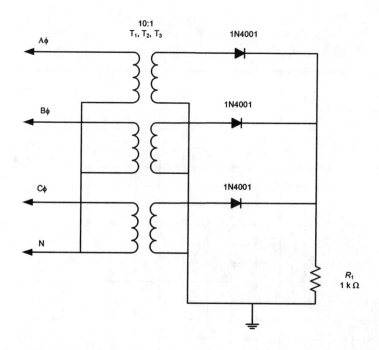

Figure 29-2 Diode load configuration for three-phase experiment

Questions

1. How did the voltage measurements differ from phase A to phase B to phase C on the secondary of the transformer?
2. What happened when the diodes were substituted in the circuit?
3. What was the frequency of the ripple in the three-phase output voltage instep 7? Why is it that frequency.

Title _____ Page No. _____

Signature _____ Witness _____ Date _____

Title _____ Page No. ___

Signature _____ Witness _____ Date _____

215

Title _____ Page No. ___

Signature _____ Witness _____ Date _____

216

Experiment 30: Frequency Response of RC and RL Circuits

Learning Objectives

After completing this laboratory experiment, you should be able to:

- Determine the transfer function and break frequencies for first-order RL and RC circuits.
- Describe and plot the frequency response for first-order RL and RC circuits.
- Determine and plot the transfer function of a first-order circuit on semi-logarithmic graphs.
- Draw and label the Bode magnitude and phase plots of the transfer function of first-order RL and RC circuits.

Background

In this experiment, you will examine the concept of frequency as a variable. The circuits in this experiment are analyzed without knowing the operating frequency. Once the frequency is known, that variable can be inserted in the analysis. Alternately the circuit response can be determined as a function of frequency.

Filter Types

The response of many circuits changes with frequency. Several of these are classified as "filters," such as "low-pass filter" (LPF), "high-pass filter" (HPF), "band-pass filter" (BPF), and so on, where the name describes the action. For example, a LPF will pass frequencies below a certain value with little attenuation, but attenuate frequencies above that value. The theoretical calculations should tell what kind of filter each circuit represents, but you really should be able to determine that by looking at the circuit and knowing how reactances vary with frequency.

The Transfer Function

The most common transfer function of a circuit is

$$H(j\omega) = \frac{\tilde{V}_o}{\tilde{V}_i}$$

H(jω) depends only on the circuit components and frequency, not on source

signal amplitudes. If $\tilde{V}_i = 1\angle 0°$, what does $\tilde{V}_o$ equal? This is a useful fact in measurements.

Logarithmic Scales

Whenever frequency response is plotted, a logarithmic frequency scale is universally used, except perhaps when only a very small frequency range is covered, in which case the linear and logarithmic scales look alike. A logarithmic scale (also called a "ratio scale") has the characteristic that equal *ratios* occupy equal spacing along the coordinate, while in a linear scale equal *differences* occupy equal spacings. On a logarithmic scale, equal spaces could also be viewed as represented by equal differences in the *exponent*, which is the reason for the term *logarithmic*.

Example: Construct a logarithmic scale with four spacings covering the range of 0.1 to 1000:

- *Method 1*: Range is 10^{-1} to 10^3, that is, a ratio of 10^4. Thus, each spacing represents a range of a factor of 10.
- *Method 2*: Total range ratio is 10,000:1. To get four equal multipliers, take the fourth root of 10,000, which is 10.

The two methods are really two views of the same calculation.

Result:

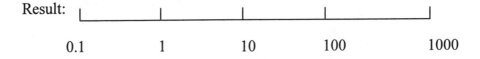

| 0.1 | 1 | 10 | 100 | 1000 |

Prelaboratory Preparation

Predict the type of filter for each circuit shown in Figure 30-1. Using a spreadsheet, generate a table and plots of the magnitude and phase of

$H(j\omega) = \tilde{V}_o / \tilde{V}_i$, both versus frequency, from 200 Hz to 200 kHz for each circuit shown in Figure 30-1. For Figure 30-1(c) go down to 20 Hz. Plot the magnitude versus frequency on semi-log graph. Plot the phase versus frequency on a linear-log graph only. Make these plots in the spreadsheet. Verify that the break frequencies are accurate in the plots.

Develop the expressions for the ratio of the open circuit (no load) output voltage to the input voltage, $\tilde{V}_o / \tilde{V}_i$, for the circuits shown in Figure 30-1. From these expressions, estimate the behavior of the output voltage as a function of frequency for each circuit by making a Bode plot. Plot them on the appropriate frequency response plots. Determine the break frequency(ies) for each circuit.

Parts List

- 200 Ω resistor
- 2.2 kΩ resistor
- 4.7 kΩ resistor
- 0.01 μF capacitor
- 100 mH inductor

Procedure

1. Measure all components. For each circuit, watch the output voltage magnitude as you rotate the frequency dial. Does it match the theoretical response, especially in any region where the output voltage changes rapidly versus frequency? If so, proceed to the next step. If not, determine what might be wrong with the components, circuit, wiring, or equipment.

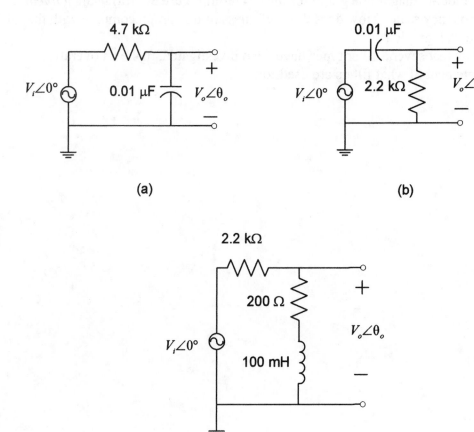

(a)

(b)

(c)

Figure 30-1   RC and RL circuits

2. Measure the output-to-input voltage ratio magnitude and phase for all three circuits over the frequency ranges used in the theoretical calculations using the " 1-2-5-10" sequence of frequencies (e.g., 10, 20, 40, 100, 200, 500, ..., Hz). Measure at more frequencies in regions where the output is changing rapidly versus frequency. *Hint*: If $\tilde{V}_i = 1\angle 0°$, what does $\tilde{V}_o$ equal?

3. Plot the measured data for each circuit on the corresponding theoretical graph. *Note*: It might be a good strategy to make the experimental plots for each circuit *while* that circuit is still set up, to avoid the need for backtracking later.

Questions

1. Compare the theoretical and measured responses and break frequencies for each circuit. Explain any significant differences.
2. Replot the magnitude graph for the circuit in Figure 30-1(a) using a *linear* frequency scale. How does the usefulness of this graph compare with that of the "log graph"?
3. Now that several filter types have been investigated, make an overall statement of what filters are used for.

Title_____ Page No. ____

Signature _____ Witness _____ Date _____

221

Title _____ Page No. ___

Signature _____ Witness _____ Date _____

Title _____ Page No. ___

Signature _____ Witness _____ Date _____

223

Title _____ | Page No. ____

Signature _____ Witness _____ Date _____

Experiment 31: Another RC Transfer Function

Learning Objectives

After completing this laboratory experiment, you should be able to:

- Develop and plot the transfer function of a first-order circuit on semi-logarithmic graphs.
- Draw and label the magnitude and phase Bode plots for the transfer function of a first-order circuit.

Background

Research on your own.

Prelaboratory Preparation

Determine the transfer function of the circuit shown in Figure 31-1. Construct a Bode plot for both magnitude and phase. You should determine the appropriate frequency range. Identify the break frequencies. Generate a frequency response plot using a spreadsheet. Verify with a circuit simulation program.

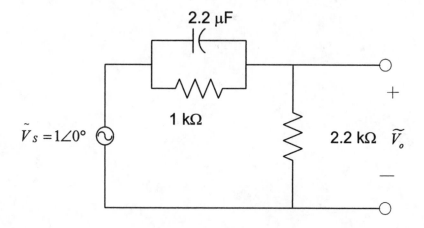

Figure 31-1 Test circuit

Parts List

- 1 kΩ resistor
- 2.2 kΩ resistor
- 2.2 μF capacitor

Procedure

Make enough measurements to verify the predicted circuit operation. You must determine the measurements that need to be taken.

Questions

1. Compare the laboratory results with the predicted and simulation results. Explain any differences.
2. Describe the behavior of this circuit. Compare it to the corresponding RC circuit of Experiment 30.

Title _____ Page No. ___

Signature _____ Witness _____ Date _____

227

Title _____ Page No. ___

Signature _____ Witness _____ Date _____

228

Title _____ Page No. __

Signature _____ Witness _____ Date _____

Title _____ Page No. ___

Signature _____ Witness _____ Date _____

Experiment 32: Another RL Transfer Function

Learning Objectives

After completing this laboratory experiment, you should be able to:

- Develop and plot the transfer function of a first-order circuit on semi-logarithmic graphs.
- Draw and label the magnitude and phase Bode plots for the transfer function of a first-order circuit.

Background

Research on your own.

Prelaboratory Preparation

Determine the transfer function for the circuit shown in Figure 32-1. Generate the Bode plots from the transfer function. Use a circuit simulation software program to verify the Bode plots. Generate a frequency response plot using a spreadsheet. Verify with a circuit simulation program.

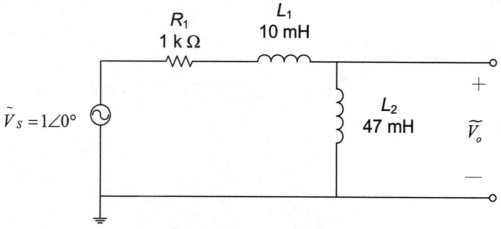

Figure 32-1 Test circuit

Parts List

- 1 kΩ resistor
- 47 mH inductor
- 10 mH inductor

Procedure

Make enough measurements to verify the predicted circuit operation. You must determine the measurements that need to be taken.

Questions

1. Compare the predicted and simulation results with the experimental results. Explain any differences.
2. Describe the behavior of this circuit. Compare it to the behavior of the RL circuit in Experiment 30.

Title _____ Page No. __

Signature_____ Witness_____ Date_____

233

Title _____ Page No. ___

Signature _____ Witness _____ Date ___

234

Title _____ Page No. __

Signature _____ Witness _____ Date _____

235

Title _____ Page No. ____

Signature _____ Witness _____ Date _____

236

Experiment 33: Series Resonance

Learning Objectives

After completing this laboratory experiment, you should be able to:

- Describe the conditions for electrical resonance.
- Determine the resonant frequency of series resonant circuits.
- Determine the quality factor of series resonant circuits.
- Determine the bandwidth of series resonant circuits.

Background

The resonant circuit has many applications in electronics. A primary use is in frequency selection. This experiment demonstrates the performance of the series resonant circuit.

Prelaboratory Preparation

Predict all phasor voltages and currents at the resonant frequency in the circuit shown in Figure 33-1. Assume the inductor has a 200 Ω internal resistance and include the 50 Ω internal resistance of the function generator in the calculations.

Determine the half-power bandwidth and frequencies. Note the simplification possible in determining these frequencies when the Q is 10 or greater. Also, calculate the voltages, currents, and impedances at these frequencies.

Use a circuit simulation software program to determine and plot the voltage magnitude and phase across the capacitor as a function of frequency. Sweep frequency from one decade below to one decade above the resonant frequency. Use 5 points/decade for the number of frequencies on a logarithmic scale.

Repeat the simulation with 100 points/decade from 0.5 f_0 to 2 f_0. What can be concluded from these two simulations?

Parts List

- 100 mH inductor
- 0.01 µF capacitor

Procedure

1. Measure all component values prior to constructing the circuit. Set the function generator to 1 V_{P-P} open circuit output. Then connect the circuit as shown in Figure 33-1. Tune the function generator until the voltage across the

capacitor is maximum. Check the open circuit voltage source output and adjust it to 1 V_P-P if necessary. Note the frequency and the voltages across the inductor and capacitor separately.

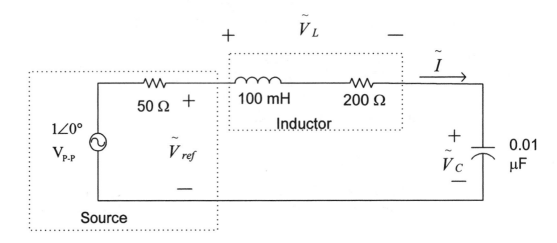

Figure 33-1 High-Q series resonant circuit

2. Adjust the generator frequency above resonance until the voltage across the capacitor drops to 70.7% of the maximum value. Check the open circuit voltage source output and adjust if necessary. Note the frequency and repeat all voltage measurements. Repeat for below resonance.

3. Measure the voltage across the capacitor for a few more frequencies (enough to verify predicted results), especially in the frequency region where the voltage is changing rapidly. Plot these experimental results onto the simulation graphs for direct comparison.

Questions

1. Calculate the experimental Q. Compare it with the theoretical Q. Were the experimental voltages across the inductor and capacitor approximately Q times the source voltage? Why or why not?
2. Did the experimental capacitor voltage response confirm theoretical predictions? Explain.
3. Compare the experimental and theoretical 3 dB bandwidth and frequencies. Explain any significant differences.

Title _____ Page No. ____

Signature _____ Witness _____ Date _____

239

Title _____ Page No. ____

Signature _____ Witness _____ Date _____

240

Title _____ Page No. _____

Signature _____ Witness _____ Date _____

241

Title _____ | Page No. ____

Signature _____ Witness _____ | Date _____

Experiment 34: Series–Parallel Resonance

Learning Objectives

After completing this laboratory experiment, you should be able to:

- Describe the conditions for electrical resonance.
- Determine the resonant frequency of series–parallel resonant circuits.
- Determine the quality factor of series–parallel resonant circuits.
- Determine the bandwidth of series–parallel resonant circuits.

Background

The resonant circuit has many applications in electronics. A primary use is in frequency selection. This experiment demonstrates the series–parallel resonant circuit performance.

Prelaboratory Preparation

The Q and 3 dB frequencies can be predicted as follows:

Calculate the resonant frequency of the circuit shown in Figure 34-1. Then convert the series inductor–resistance combination into its parallel equivalent. Convert the series voltage source/100 kΩ resistor into an equivalent current source in parallel with the resistance (source conversion). Calculate the circuit Q for this equivalent parallel resonant circuit (the equivalent circuit is valid *only* at the resonant frequency). The Q should be greater than 10. Then calculate the half-power bandwidth and frequencies.

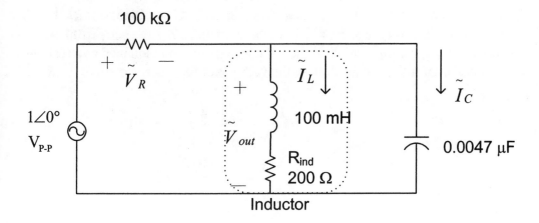

Figure 34-1 High-Q series–parallel resonant circuit

Find all voltages and currents at the resonant and 3 dB frequencies in the circuit shown in Figure 34-1. Assume the inductor has a 200 Ω resistance (the 50 Ω function generator resistance is negligible compared to the 100 kΩ resistor). Note the magnitude of the inductor and capacitor currents with respect to the source current.

Use circuit simulation to determine and plot the voltage magnitude and phase across the inductor/capacitor as a function of frequency. Sweep frequency from one decade below to one decade above the resonant frequency. Use 5 points/decade for the number of frequencies on a logarithmic frequency scale.

Repeat the circuit simulation with 100 points/decade from $0.5 f_0$ to $2 f_0$. What do you conclude about the circuit simulation when you compare both simulation plots?

Parts List

- $0.0047 \mu F$ capacitor
- 100 mH inductor
- 100 kΩ resistor

Procedure

1. Measure all component values before constructing the circuit. Set the function generator to a 1 V_{P-P} open circuit output. Then connect the circuit shown in Figure 34-1. Tune the function generator until the voltage across the capacitor is at maximum. Check the open circuit voltage source output and adjust if necessary. Note the frequency and the voltage (both magnitude and phase) across the capacitor. Measure the voltage magnitude across the 100 kΩ resistor.

2. Calculate the equivalent resistance of the parallel LC combination (R_P) by comparing the voltage across R_P with the voltage across the series 100 kΩ resistor. *Suggestion*: Think of the LC combination at resonance as a resistor R_P, as shown in Figure 34-2. Use the voltage divider rule and solve for R_P.

$$V_{out} = 1\angle 0° \left(\frac{R_P}{100 \text{ k}\Omega + R_P} \right)$$

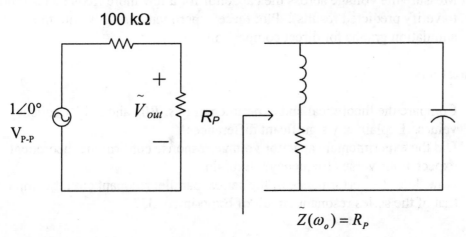

$$\tilde{Z}(\omega_o) = R_P$$

Figure 34-2 Equivalent resistance of the parallel resonant circuit at resonance

3. Calculate an experimental Q of the entire circuit by placing R_P in parallel with the source equivalent resistance of the series voltage source, that is, the 100 kΩ resistor, as shown in Figure 34-3.

$$Q = R_T \omega_0 C$$

where $R_T = 100 \text{ k}\Omega \parallel R_P$.

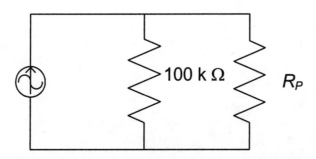

Figure 34-3 Parallel equivalent circuit of the series–parallel resonant circuit at resonance

4. Decrease the function generator frequency until the voltage across the resonant circuit has decreased to 70.7% of its maximum value. Check open-circuit voltage source output. Note the frequency and measure the voltage phase. Repeat for the 3 dB point above resonance. Calculate another experimental Q of the circuit from these measurements.

5. Measure the voltage across the capacitor for a few more frequencies (enough to verify predicted results). Plot these experimental results onto the simulation graphs for direct comparison.

Questions

1. Compare the theoretical and experimental Q's, BW, and 3 dB frequency values. Explain any significant differences.
2. Did the experimental capacitor voltage response confirm the theoretical expectations versus frequency? Explain.
3. How does the performance of the series–parallel resonant circuit compare to that of the series resonant circuit of Experiment 33?

Title _____ Page No. ___

Signature _____ Witness _____ Date _____

Title _____ Page No. _____

Signature _____ Witness _____ Date _____

248

Title _____ Page No. ____

Signature _____ Witness _____ Date _____

249

Title _____ Page No. ___

Signature _____ Witness _____ Date _____

Appendix A

Sample Laboratory Report

The names utilized in this laboratory report are fictitious and in no way represent any known persons.

Sample Title Page

[COLLEGE/UNIVERSITY NAME]

[DEPARTMENT NAME]

ELECTRICAL POWER IN DC CIRCUITS

(course number)   Experiment 0

Submitted to:	Prof. T. Cher
Submitted by:	Stu Dent
Date:	09/10/2002
Lab Partner:	P. Art Ner

Sample Table of Contents page

TABLE OF CONTENTS

(Double spacing of text *must* be used in formal laboratory reports.)

{*Note*: Independent research should be part of any report. Appropriate references should be used. Verify with your instructor the correct method of referencing sources for your institution.}

INTRODUCTION

A DC circuit with voltage applied and current flowing dissipates power. It is important to know the power that will be dissipated in electrical components so that their power ratings can be properly specified. Hence, it is the purpose of this experiment to predict, experimentally determine, and compare electrical powers in a DC circuit in order to determine if the DC power equations accurately predict the actual power in a DC circuit.

THEORETICAL SOLUTION

The electrical power in a DC circuit can be predicted using the equation[1]

$$P = VI = \frac{V^2}{R} = I^2 R$$

where: P = power in watts (W),

V = voltage in volts (V),

I = current in amperes (A), and

R = resistance in ohms (Ω).

For the circuit in Figure 1, the predicted current for a 5 V supply voltage is:

$$I = \frac{V}{R} = \frac{5\ \text{V}}{2\ \text{k}\Omega} = 2.5\ \text{mA}$$

The predicted power can be found from the voltage and current:

$$P = VI = (5\ \text{V})(2.5\ \text{mA}) = 12.5\ \text{mW}$$

---

[1] Strangeway, et al, *Contemporary Electric Circuits: Insights and Analysis*, Prentice-Hall, 2003.

The currents and powers for the other power supply voltages are similarly determined and the results are reported in Table 1 of the Results section.

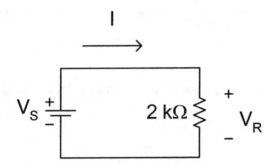

Figure 1. DC circuit for power determination

RESULTS

Procedure

The circuit in Figure 1 is built. An ammeter is placed in series and a voltmeter is placed in parallel with the resistor. The voltage and current are measured for each power supply voltage listed in Table 1.

Data

Table 1. Theoretical and measured results for the circuit in Figure 1

Nominal Supply Voltage (V)	Measured Voltage (V)	Theoretical Current (mA)	Measured Current (mA)	Theoretical Power (mW)	Actual Power (mW)
5.0	5.1	2.5	2.6	12.5	13.3
10.0	10.0	5.0	4.9	50.0	49.0
15.0	14.8	7.5	7.4	112	110
20.0	20.1	10.0	10.3	200	207
25.0	24.9	12.5	12.6	312	314

$R = 2.02$ kΩ (measured on the ohmmeter)

<u>Sample Calculation</u>

$$P_{acutal} = V_{meas} \; I_{meas} = (5.1\,\text{V})\,(2.6\,\text{mA}) = 13.3\,\text{mW}$$

ANALYSIS OF RESULTS

<u>Graphs</u>:

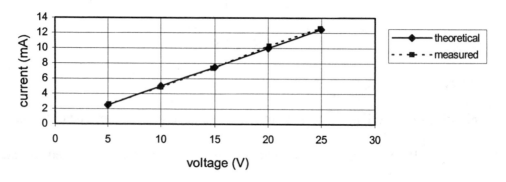

Figure 2. Current vs. voltage plot for the circuit in Figure 1

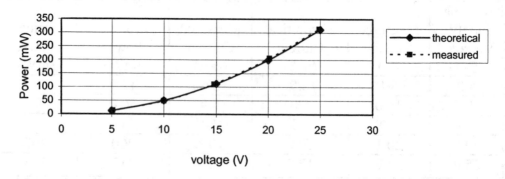

Figure 3. Power vs. voltage graph for the circuit in Figure 1

Discussion

The measured current and power results confirm the trends in the theoretical solution. The current rises linearly as the source voltage increases but the power does not. The power curve looks like a parabola. This is due to the voltage-squared factor in the power equation.

Error Analysis

The maximum percentage difference between the theoretical and measured powers is 6.4 percent for the 5 V case. This is within the expected range considering the resistor tolerance is 5 percent. If the lowermost resistance value is utilized in the theoretical power calculation:

$$P = \frac{V^2}{R} = \frac{5.1^2}{(2000)(0.95)} = 13.7 \text{ mW}$$

Hence, the measured power of 13.3 mW is less than the maximum theoretical power with the resistance at the lowest edge of the tolerance range. This also does not consider any additional error contributed by the voltmeter and the ammeter. Thus, the data is considered reliable.

CONCLUSION

The validity of the equation to predict the power in a DC circuit, $P = VI$, was tested in this experiment. The validity of this equation has been verified for a single-resistor circuit. In this experiment, it was also found that the current rises linearly and the power rises in a parabolic manner as the source voltage increases in a single-resistor DC circuit. Overall, this experiment has verified that power can be accurately predicted in a DC circuit.

Appendix B

Sample Engineering Notebook Entry

Note on recording in an engineering notebook:

The pages that follow contain typical entries in an engineering notebook. The title of the experiment would be at the top of the first page of the experiment in an engineering notebook. Most of the text would be handwritten, but the table could be generated from a spreadsheet and the graphs could be the output of a computer software package such as a spreadsheet. Any computer-generated material must be permanently attached to the engineering notebook pages.

The names utilized in this laboratory report are fictitious and in no way represent any known persons.

ELECTRICAL POWER IN DC CIRCUITS   EXPERIMENT 6

INTRODUCTION

A DC circuit with voltage applied and current flowing will dissipate power. It is important to know the power will be dissipated in electrical components so that their power rating can be properly specified. Hence, it is the purpose of this experiment to predict, experimentally determine, and compare electrical powers in a DC circuit in order to determine if the DC power equations accurately predict the actual power in the circuit.

THEORETICAL SOLUTION

The electrical power in a DC circuit can be predicted using the equation

$$P = VI = \frac{V^2}{R} = I^2 R$$

where
P = Power in Watts (W),
V = voltage in Volts (V),
I = current in Amperes (A), and
R = resistance in Ohms (Ω).

Stu Dent 9/10/2002

4

The ~~ci~~ [SD 9/10/2002] current for the circuit in Figure 1 can be predicted using:

$$I = \frac{V}{R}$$

For the 5V source:

$$I = \frac{V}{R} = \frac{5V}{2k\Omega} = 2.5mA$$

The power can be predicted from the voltage and the current.

$$P = VI = (5V)(2.5mA) = 12.5mW$$

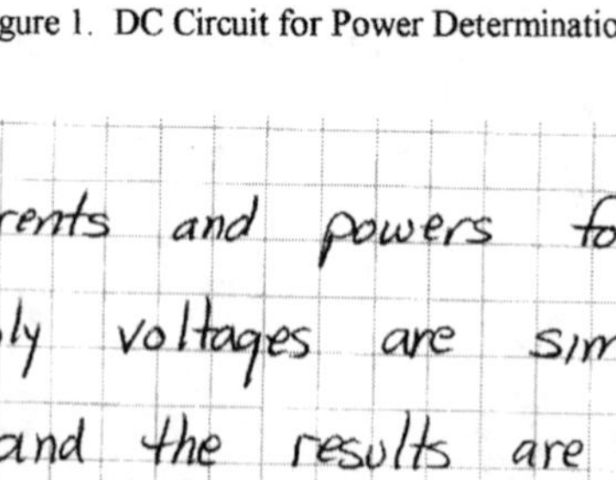

Figure 1. DC Circuit for Power Determination

The currents and powers for the other power supply voltages are similarly ~~calculated~~ [SD 9/10/2002] determined and the results are reported in Table 1 of the Results section.

Stu Dent 9/10/2002

The circuits were simulated using PSPICE.

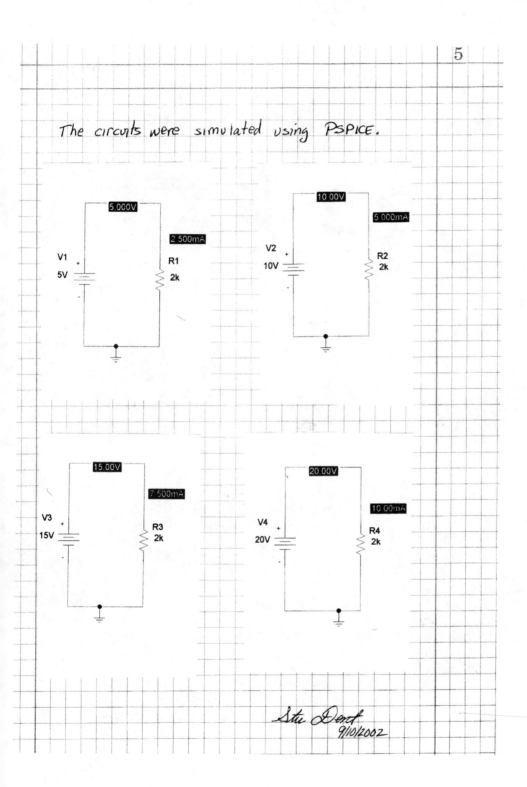

Stu Dent
9/10/2002

25.00V

12.50mA

V5
25V

R5
2k

The simulated results
are compared in
the Analysis of Results
section.

RESULTS

Procedure:

The circuit in Figure 1 was built. An
ammeter was placed in series and a ~~volta~~
voltmeter was placed in parallel with the
resistor. The voltage and current were
measured for each power supply voltage listed
in Table 1.

SD 9/10/2002

Stu Dent
9/10/2002

Table 1. Theoretical and Measured Results for the Circuit in Figure 1

Nominal Supply Voltage (V)	Measured Voltage (V)	Theoretical Current (mA)	Measured Current (mA)	Theoretical Power (mW)	Actual Power (mW)
5.0	5.1	2.5	2.6	12.5	13.3
10.0	10.0	5.0	4.9	50.0	49.0
15.0	14.8	7.5	7.4	112	110
20.0	20.1	10.0	10.3	200	207
25.0	24.9	12.5	12.6	312	314

R = 2.02 kΩ (measured on the ohmmeter)

Sample Calculation:

SD 9/10/2002

$$P_{actual} = V_{meas} \, I_{meas} = \frac{(4.9V)(2.6mA)}{(5.1V)} = 13.3 \, mW$$

ANALYSIS OF RESULTS

Graphs:

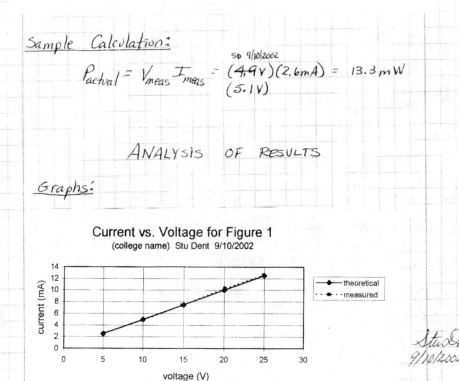

Current vs. Voltage for Figure 1
(college name) Stu Dent 9/10/2002

Figure 2. Current vs. Voltage Plot for the Circuit in Figure 1

Stu Dent
9/10/2002

Power vs. Voltage for Figure 1
(college name) Stu Dent 9/10/2002

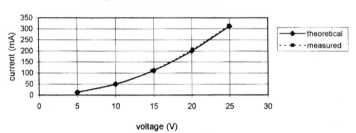

Figure 3. Power vs. Voltage Graph for the Circuit in Figure 1

Discussion:

The measured current and power confirm the theoretical solution. The current rise linearly as the source voltage increases SD 9/10/2002 increases but the power does not. The power curve looks like a parabola. This is due to the square SD 9/10/2002 voltage-square factor in the power equation. The measured current also confirms the simulation solution.

Stu Dent
9/10/2002

<u>Error Analysis:</u>

The maximum percentage difference between the theoretical and measured powers is 6.4 percent for the 5V case. This is within the expected range considering the resistor tolerance is five percent. If the lowermost resistance value is utilized in the theoretical power calculation:

$$P = \frac{V^2}{R} = \frac{(5.1)^2}{(2000)(0.95)} = 13.7\,mW$$

Hence, the measured power of 13.3 mW is less than the maximum theoretical power with the resistance at the lowest edge of the tolerance range. This ~~is~~ _{SD 9/10/2002} also does not consider any additional error contributed by the voltmeter and ammeter. Thus, the data is considered reliable.

Stu Dent
9/10/2002

SD 9/10/2002

~~CONT'D~~

CONCLUSION

The validity of the equation to predict the power in a DC circuit, $P = VI$, was tested in this experiment. The validity of this equation has been verified for a single resistor circuit. In this experiment, it was found that the current rises linearly and the power rises in a parabolic manner as the source voltage increases in a single resistor DC circuit. Overall, this experiment has verified that power can be accurately predicted in a DC circuit.

Stu Dent
9/10/2002

Part Ner
9/10/2002

2n3

For information about additional
Prentice Hall Electronics Technology
resources, visit www.prenhall.com

ISBN 0-13-111560-X

90000

PEARSON
Prentice
Hall

StudentAid.ed.gov
FUNDING YOUR FUTURE.

9 780131 115606